Teen Nonfiction - PORTR-HEB
001.42 CURRI
Currie, Stephen
Biased science
33410018221608   09-26-2022

Biased Science

Stephen Currie

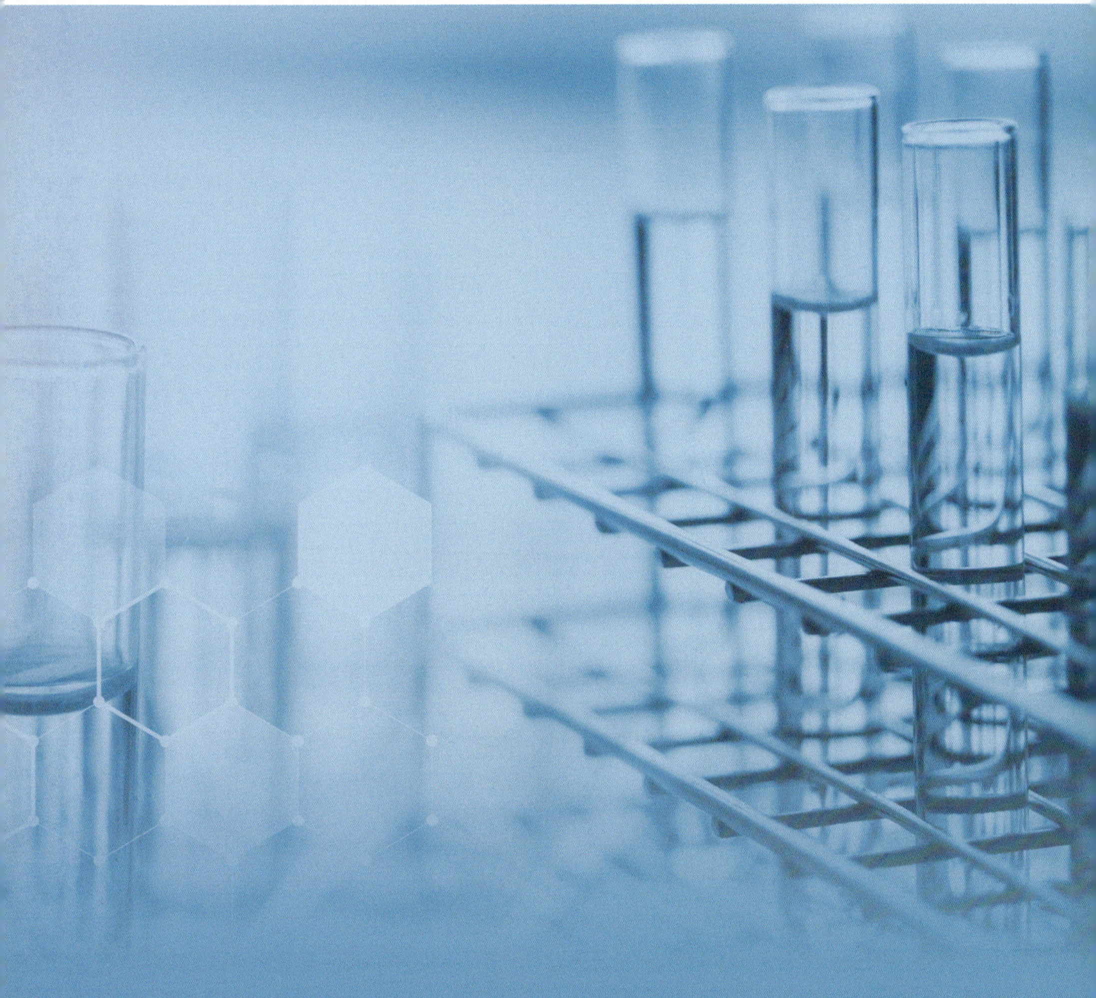

San Diego, CA

About the Author

Stephen Currie is the author of dozens of books for young people, including many for ReferencePoint Press. He has also worked as a writer and editor of textbooks, teacher guides, and other educational materials, and he has taught grades ranging from kindergarten to college. He lives with his family in New York's Hudson Valley.

© 2023 ReferencePoint Press, Inc.
Printed in the United States

For more information, contact:
ReferencePoint Press, Inc.
PO Box 27779
San Diego, CA 92198
www.ReferencePointPress.com

ALL RIGHTS RESERVED.
No part of this work covered by the copyright hereon may be reproduced or used in any form or by any means—graphic, electronic, or mechanical, including photocopying, recording, taping, web distribution, or information storage retrieval systems—without the written permission of the publisher.

Picture Credits:
Cover: kurhan/Shutterstock Images (top);
    peterschreiber.media/Shutterstock Images (bottom)

 6: Chronicle/Alamy Stock Photo
10: Associated Press/National Archives
13: SPL/Science Source
14: Pictorial Press Ltd/Alamy Stock Photo
20: PA Images/Alamy Stock Photo
23: Associated Press

26: vladimir salman/Shutterstock
30: Monkey Business Images/Shutterstock Images
33: Chaay_Tee/Shutterstock Images
36: Imaginechina Limited/Alamy Stock Photo
40: mimagephotography/Shutterstock Images
42: Operation 2021/Alamy Stock Photo
45: Mark Scheuern/Alamy Stock Photo
50: Tong_stocker/Shutterstock Images
53: Arne Beruldsen/Shutterstock Images
55: DC Studio/Shutterstock

LIBRARY OF CONGRESS CATALOGING-IN-PUBLICATION DATA

Names: Currie, Stephen, author.
Title: Biased Science / Stephen Currie.
Description: San Diego, CA : ReferencePoint Press, 2022. |
  | Includes bibliographical references
    and index.
Identifiers: LCCN 2022014074 (print)| ISBN
  9781678202323 (library binding) | ISBN 9781678202330 (ebook)
Subjects: LCSH: scientific research | Medical
  ethics--Juvenile literature. | Medical research
  Communicable diseases--Juvenile literature.

CONTENTS

Introduction 4
Scientific Bias

Chapter One 8
The Tuskegee Experiment and Henrietta Lacks

Chapter Two 18
Scientific Fraud

Chapter Three 28
Artificial Intelligence and Algorithms

Chapter Four 38
Sexism and Design

Chapter Five 48
Medicine and Medical Trials

Source Notes 57
For Further Research 60
Index 62

INTRODUCTION

Scientific Bias

Science is a central part of the human quest for understanding and knowledge about the world around us. Life sciences such as biology and zoology help humans make sense of the living world and the way various species interact. Through earth sciences such as geology and meteorology, we learn about how the earth was formed and how it can best be protected. Physics, chemistry, and other physical sciences teach us about matter and how it behaves throughout the universe. Scientists often see their work as seeking truth on a grand scale. Moreover, they like to view what they do as dispassionate and impervious to error. "A scientific theory," writes Nobel Prize–winning scientist Steven Weinberg, "is culture-free and permanent."[1]

But in fact, science is not at all permanent. On the contrary, scientific thinking has changed many times and in many ways over the years. Until just a few hundred years ago, after all, scientists believed that the sun revolved around the earth. Germ theory, which explains how certain diseases are transmitted, is less than two centuries old. So is the theory of evolution. The notion of continental drift—that landmasses have moved around the globe over the years instead of remaining in a single fixed position—was not developed until the middle of the 1900s. Far from being stable and enduring, science is ever-changing; we discard and reshape old ideas as we learn more about what the universe is actually like.

Neither is science culture-free. For years it has largely been the domain of White men from Europe and North America, men who carried with them the biases and prejudices of the times and places where they lived. The renowned scientist Charles Darwin, for example, firmly believed that women were not as fully developed as men. "Man attain[s] to a higher eminence, in whatever he takes up, than woman can attain," he wrote. "Thus man has ultimately become superior to woman."[2] Darwin did no formal scientific experiments to prove his claim. No doubt he did not believe he needed to, for to him the statement was obviously true. In his time and place—Victorian England—the inferiority of women was a widely accepted view.

> "Man has ultimately become superior to woman."[2]
>
> —Scientist Charles Darwin

Similarly, scientific consensus for generations held that Black people were inferior to White people both morally and intellectually. Notions of White superiority were often couched in a veneer of scientific terminology—and supposed scientific fact. For example, scientists of the eighteenth and nineteenth centuries asserted that Black people had smaller brains than White people and therefore were of less intelligence. In the early twentieth century, scientists reasoned that African Americans were predisposed toward criminal behavior because there were so many Black people in American prisons. Of course, this analysis failed to account for much more significant factors such as poverty and a biased criminal justice system. But many racist Whites of the period were happy to use the language of science to argue their points.

Ethics, Fraud, and Prejudice

Since the nineteenth century our scientific understanding of the world has improved greatly. As a society, we also have developed a much keener ability to recognize the biases and prejudices of the scientists who dominated the scientific landscape in the past. But that is not to say that we have reached Weinberg's goal and now routinely do science in a way that is both permanent and culture free. Quite the reverse; science today remains subject to all kinds of biases, hidden and otherwise.

Some of these biases involve ethics, or the set of moral behaviors that go along with scientific study. Over time scientists and others have come up with ethical codes for scientific research. If at all possible, for example, participation in a scientific study should never harm volunteers. In addition, it is essential that the

Scientist Charles Darwin, pictured, believed that it was a scientific fact that woman were inferior to men.

subjects of scientific research understand the ramifications of a study—and be given the right to opt out if they choose to do so. These may not sound like radical ideas. Yet often they have been widely ignored, to the detriment of study participants—and to the detriment of scientists and the scientific process as well.

Scientific fraud is another way in which science is not nearly as focused on the truth as it may appear. There are often strong pressures for scientists to find certain results in their research. In particular, scientists may feel pushed to come up with certain results based on political or financial criteria. In some cases these considerations outweigh the goal of unbiased inquiry into a given topic. Once again, the image of science as neutral and balanced suffers. And racism and sexism still play a significant role in scientific endeavors such as making medical diagnoses, developing product designs, and creating algorithms to be used with artificial intelligence.

Ideally, science would indeed be focused entirely on facts, truth, and objectivity. But the reality is different. Science cannot be separated from the human experience. As long as science is a human endeavor, it will carry with it the biases of society. It is always a good idea to examine scientific information carefully to see where these biases may be hiding.

CHAPTER ONE

The Tuskegee Experiment and Henrietta Lacks

In 1932 a government agency called the US Public Health Service (PHS) organized a medical study involving syphilis, a sexually transmitted disease that can kill those who have it. The study centered on the small Alabama town of Tuskegee and was carried out at a clinic on the campus of a Black college called the Tuskegee Institute. The PHS recruited 600 men from the area to take part in the study. Of these men, 399 had syphilis. All were African American; virtually all were poor and uneducated. Study participants were offered medical care, rides to and from the clinic, meals on days when they were being given examinations, and burial stipends to be paid to their survivors should any of the participants die during the study.

Everyone involved in planning the study knew that the research was not about curing the men who had the disease. At the time, there were no truly effective treatments for syphilis. The only available protocols involved poisonous substances such as arsenic and mercury. These treatments were painful and led to unpleasant side effects. As one expert put it, the drugs offered "more potential harm for the patient than potential benefit."[3] Nor did researchers aim to find a wonder drug that would provide a cure. Rather, researchers were simply interested in tracking the course of the disease in those it afflicted. There was no expectation that the men who had syphilis would recover.

But that was not what the men recruited for the study were told. Instead, the PHS promised prospective volunteers that they would receive appropriate treatment for what ailed them. Recruitment flyers explained that all volunteer subjects would be given a physical examination. "After [the examination] is finished," the flyer continued, "you will be given a special treatment if it's believed that you are in a condition to stand it." The flyer ended with a veiled threat, printed in all capital letters: "REMEMBER THIS IS YOUR LAST CHANCE FOR SPECIAL FREE TREATMENT."[4]

> "They just kept saying I had the bad blood. They never mentioned syphilis to me, not even once."[5]
>
> —Tuskegee experiment test subject Charles Pollard

Moreover, evidence strongly suggests that the men were not told that the study was about syphilis. Charles Pollard, one study participant, told later interviewers that he had been recruited because he had what researchers called "bad blood"—a catch-all term within the local Black community for ailments both serious and benign. "They just kept saying I had the bad blood," Pollard reported years later. "They never mentioned syphilis to me, not even once."[5] And while some study leaders disputed Pollard's recollections, other doctors involved in the research backed his account.

Tuskegee and Medical Ethics

From a perspective of medical ethics, the Tuskegee study was severely flawed. Principles of medical ethics state that studies should only be performed on people who have given informed consent—that is, people who know what the study is about and what their role in it will be. The Tuskegee syphilis study failed

This photo from the 1950s shows men who were included in the Tuskegee, Alabama, syphilis study. From a perspective of medical ethics, the Tuskegee study was egregiously unethical.

to provide enough information to subjects to ensure that they could make an informed decision about whether to participate. Moreover, by not identifying syphilis as the focus of the study, the researchers helped spread the disease through the community. Study participants did not know that they could infect their wives—and their unborn children, since syphilis can be transmitted to a fetus if the mother has the disease—and so dozens of women and children became syphilitic as well.

Matters grew considerably worse in the early 1940s, when penicillin, an antibiotic, was used to treat syphilis for the first time. It soon became evident that penicillin cured the disease and carried few negative side effects. But those in charge of the Tuskegee study decided not to dose the men in their care with penicillin. Instead, the study continued as before, with doctors recording the inevitable health declines of the subjects who had syphilis. "I hope that the availability of antibiotics has not interfered too much with this project,"[6] commented Raymond Vonderlehr, one of the study's directors, in 1952—making the goal of the experiment clear.

No one knows how long the study might have continued. But in 1972 reporter Jean Heller published a story about the study. Heller's account was met with shock and horror. For decades a government health agency—an agency dedicated to promoting health—had lied to these Black men about their medical condition and denied them treatments known to be effective in fighting their disease. Though study organizers initially tried to downplay the immorality of the experiment, most observers remained aghast. Journalist Harry Reasoner wondered how the PHS could be "only mildly uncomfortable" with using "human beings as laboratory animals in a long and inefficient study of how long it takes syphilis to kill someone."[7]

The Tuskegee study ended soon after Heller's account appeared. But the effects lingered. The federal government put together a panel of experts to review the study; unsurprisingly, the panel concluded that the research was "ethically unjustified."[8] A lawyer sued the government on behalf of the men in the study and their families, eventually settling for more than $10 million. In 1997 then-president Bill Clinton apologized to the few surviving research subjects and to the families of all who had participated. "We can stop turning our heads away," Clinton said. "We can look at you in the eye and finally say on behalf of the American people, what the United States government did was shameful, and I am sorry."[9]

Henrietta Lacks

The Tuskegee research project is unfortunately far from the only example of an unethical medical study. Another example is the case of Henrietta Lacks. Born in Virginia in 1920, Lacks was African American and poor. One of ten children, she was raised mainly by her grandfather and attended school through only the sixth or seventh grade. After she was married, she moved to Baltimore, Maryland. According to friends and relatives, Lacks enjoyed cooking, adored her five children, and loved to dance. As one of her cousins put it, "We'd just get out there [on the dance floor] and shake and turn around and all like that."[10]

Denying Treatment in Tuskegee

As time passed, doctors both in and outside of the Tuskegee study realized that they were not permitted to treat the subjects of the experiment in any meaningful way. Since the purpose of the study was to examine the effects of syphilis, providing medication to combat the men's symptoms was forbidden, even for doctors not involved in the study, because it might interfere with the study results.

Dr. Reginald James, who worked with syphilis patients as part of public health programs in the Tuskegee area from 1939 to 1941, remembered a nurse named Eunice Rivers who told him to ignore the medical needs of study subjects. "He's under study and not to be treated," James recalled Rivers saying when he wanted to offer treatment to the men. James did not have much effective medication to offer; still, he found Rivers's attitude deeply distressing.

Similarly, after penicillin became available, the men were not told that it might be helpful, which might have allowed them to leave the study and seek effective medical care elsewhere. Throughout the experiment, researchers made it abundantly evident that the validity of their study took precedence over the health of their subjects.

Quoted in James H. Jones, *Bad Blood*. New York: Free Press, 1981, p. 6.

In 1951, however, Lacks began experiencing acute abdominal pain and bleeding. She went to Johns Hopkins Hospital in Baltimore—one of the most prestigious hospitals in the country, and one of the few in the Baltimore area that would treat Black people—where she was told that she had cervical cancer. Lacks told her family members that there was nothing to worry about. "Doctor's gonna fix me right up,"[11] she assured her husband and children. She was treated with radium, a radioactive element that despite serious side effects was known to kill cancer cells. However, Lacks died on October 4, 1951, less than a year after her diagnosis.

But even though Lacks was dead, some of her cells were still alive. During her visits to Johns Hopkins, doctors had taken samples of Lacks's malignant cells and given them to medical researcher George Gey. George and his wife, Margaret Gey, were attempting to grow cancerous cells outside the human body. If they could find a way to do so, they would be able to run experiments on the cells without the possibility of inflicting harm on living cancer patients. Assuming that the malignant cells could be made to grow and reproduce outside the body as they did inside it, then the Geys could try various treatments on the cells, learn about the causes of cancer, and perhaps even develop cures for the disease.

Though the Geys had been unsuccessful in growing cells to this point, the cells taken from Henrietta Lacks proved different. The cell sample doubled, then doubled again. Lab assistants filled one test tube after another with Lacks's cells. "Spreading like crabgrass!"[12] remarked Margaret when she saw what was happening. The cells became a cell line, a term referring to a group of cells that proliferate forever under the right circumstances. The term *immortal* is also sometimes used to describe cell lines like these. The Geys named this immortal cell line the HeLa line, using the first two letters of Lacks's first and last names. They also made the HeLa cells available to other researchers.

As the Geys hoped and predicted, these immortal cells turned out to be invaluable to medical studies. As early as 1953,

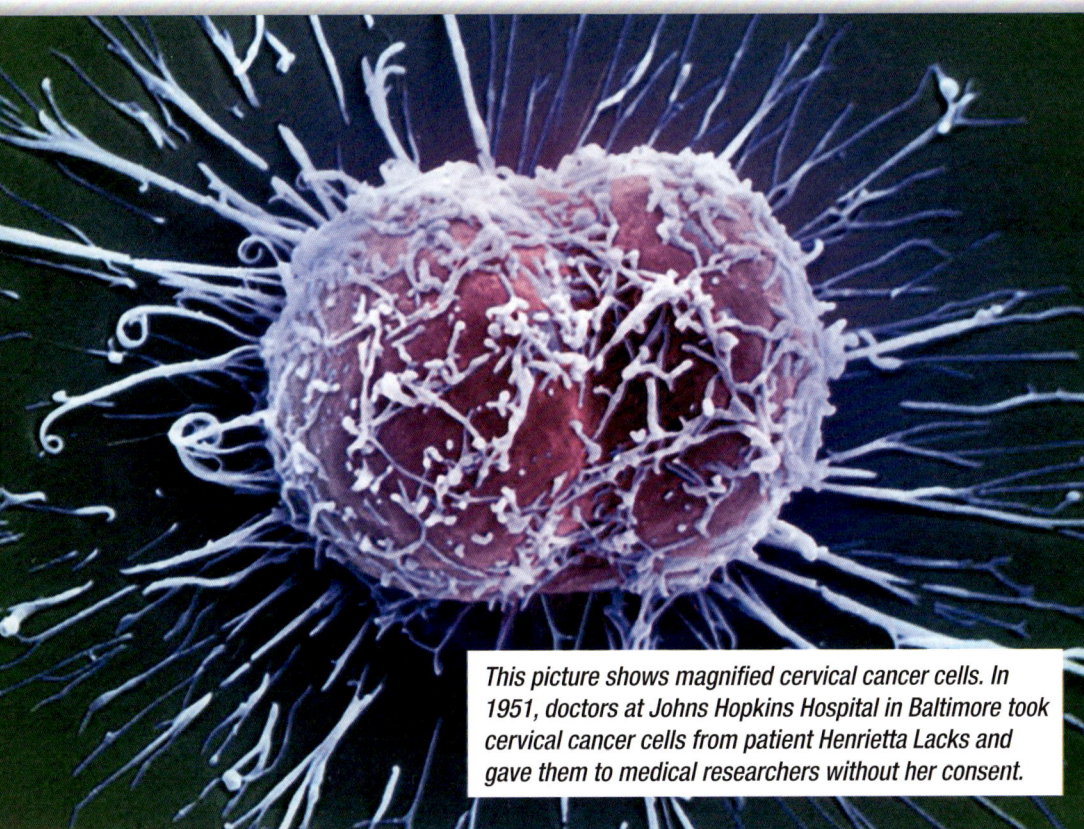

This picture shows magnified cervical cancer cells. In 1951, doctors at Johns Hopkins Hospital in Baltimore took cervical cancer cells from patient Henrietta Lacks and gave them to medical researchers without her consent.

The cervical cancer cells taken from Henrietta Lacks turned out to be invaluable to medical studies. For example, Jonas Salk—pictured—used them to develop a vaccine for polio.

a physician named Jonas Salk used Lacks's cells to develop a vaccine for a serious disease called polio. HeLa cells have also been instrumental in the fight against human immunodeficiency virus (HIV), cancers—including cervical cancer, the disease that ultimately killed Lacks—and once common childhood diseases such as measles and mumps. Even today HeLa cells are still in frequent use in a variety of medical studies. By 2009, according to author Rebecca Skloot, at least sixty thousand scientific papers had been published based at least in part on research using the HeLa cell line, with hundreds more appearing each month.

Ethical Questions

Though Lacks's cells have unquestionably been good for the world, their use raises undeniable ethical questions. Primary among these, as in the case of the Tuskegee study, is the issue of informed consent. Doctors at the hospital did have Lacks sign a release form giving them the right to perform any surgical procedure on her that they thought were necessary. But as the American Medical Association puts it, "Informed consent is more than simply getting a patient to sign a written consent form. It is a process of communication between a patient and a physician."[13]

In the case of Henrietta Lacks, though, that communication never took place. No one at Johns Hopkins asked Lacks for her permission to harvest cells from her body; no one told her that those cells would be used in a medical study. Nor did anyone tell Lacks's family what George and Margaret Gey planned to do with the cells. Thus, Lacks's husband and children had no opportunity to raise objections to the use of the cells in research or to place limitations on how those cells would be used. Nor did they benefit financially from the arrangement—as did many of the scientists who used her cells, along with the companies they worked for. In sum, Lacks's physicians largely ignored medical ethics in their treatment of her and in the harvesting of her cells.

> "Informed consent is more than simply getting a patient to sign a written consent form. It is a process of communication between a patient and a physician."[13]
>
> —American Medical Association

The principle of informed consent was less carefully observed in 1951 than it is today. Evidence shows that doctors at Johns Hopkins and elsewhere often used their patients as subjects in medical experiments without telling them what was going on or giving them a chance to object. That was especially true of poor African American patients. Still, even judging by the standards of 1951, it is apparent that doctors and researchers should have

Harry Bailey's Experiments

The Tuskegee experiment is the most famous example in American history of Black people being mistreated in the name of science, but there are others. The case of Harry Bailey, an Australian psychiatrist who came to the United States in the 1950s, is one example. Bailey worked for Tulane University in Louisiana, and indirectly for the Central Intelligence Agency (CIA). It was the height of the Cold War between the United States and the Soviet Union, and the CIA was interested in exploring mind control techniques, which it enlisted Bailey to research. All of Bailey's research subjects were Black, many of them prison inmates or hospital patients. There is no evidence that Bailey obtained informed consent from any of his subjects for experiments that included the administration of mind-altering drugs such as LSD and a stupor-inducing drug known as bulbocapnine. He also performed electric stimulation of various parts of subjects' brains. As Harriet A. Washington writes in her book *Medical Apartheid*, Bailey implanted electrodes into his subjects' brains and then sent charges into the areas of the brain responsible for feelings. "By stimulating these areas," Washington writes, "Bailey evoked pleasure, pain, joy, anger . . . and other powerful emotions in his black subjects at will."

Harriet A. Washington, *Medical Apartheid*. New York: Doubleday, 2006, p. 356.

given Lacks more information about her condition—and should have let her decide whether to allow her cells to be used for medical research.

The Need for Ethics

On one hand, the Tuskegee study and the case of Henrietta Lacks have little in common. The Tuskegee study had no medical merit and led to no cure or treatment for syphilis. In contrast, Lacks's cells have played a major role in developing vaccines, treatments, and cures for a wide variety of medical conditions. The subjects of the Tuskegee study were knowingly denied effective medical care, while the treatments Lacks received were entirely appropriate for a woman with cervical cancer at the time. The men of Tuskegee were deliberately and repeatedly lied to about almost everything connected with their condition; that was not the case for Lacks.

But the two cases are linked by the fact that the scientists involved paid little attention to medical ethics in designing and carrying out their studies. The Tuskegee study should never have taken place; Lacks should have been asked her permission for her cells to be studied. The cases are also connected because both studies involved poor African Americans with little formal education. It is difficult to imagine the Tuskegee research being carried out in the same way with a group of White college graduates. And as author Ta-Nehisi Coates writes about Lacks, "When you understand the incredible web of racism which gripped this country in 1951, it becomes very hard to look at any black person living in that time and say 'this would have happened exactly the same way to everyone.' Racism changes everything."[14]

Science—and medicine is a part of science—is biased in many ways. As the cases of Henrietta Lacks and the men of Tuskegee demonstrate, it is often prejudiced against African Americans and the poor. Given what happened in these two cases, it is perhaps no surprise that Black Americans are less trusting of scientists today than are White Americans; as one study notes, "There's a 14-point gap between the shares of White and Black adults who say they have a great deal of confidence in scientists."[15] In both Tuskegee and Baltimore, researchers were too interested in their own goals and not interested enough in the people they were using to achieve those goals. When doctors and researchers violate or ignore medical ethics, the reputation of science as a fair-minded and dispassionate search for truth is damaged.

CHAPTER TWO

Scientific Fraud

Few public health measures over the years have been as effective as vaccination. Widely used today to combat the spread of serious diseases such as influenza, polio, and COVID-19, vaccines were first used in the 1700s to keep people from contracting a deadly disease called smallpox. At the time millions around the globe died from this disease every year. Many more survived the disease but were badly disfigured. Thanks to vaccination, smallpox was eradicated in the late 1970s.

Other vaccines may not be quite as effective as the smallpox vaccine, but their track record has nonetheless been impressive. Once, nearly every American child got chicken pox, a relatively mild viral infection that nonetheless caused more than one hundred deaths in the United States per year. Today, in contrast, fewer than twenty Americans die of chicken pox annually, according to the Centers for Disease Control and Prevention (CDC), and cases have fallen from more than 4 million to less than 350,000 each year. Influenza vaccines, similarly, have been credited with keeping about 7.5 million Americans from getting the flu every year. And the Commonwealth Fund, a health care organization, estimates that as of late 2021, the COVID-19 vaccine had prevented over 1 million deaths and 10 million hospitalizations.

Despite their well-documented success, there has been much controversy related to vaccines over the years. Overwhelming evidence shows that the COVID-19 vaccine protects most people from severe illness and death. Yet, it has come under fire from people convinced that the vaccine should not be given to anyone. Many of these people believe that the threat of COVID-19 is severely overblown and assert that the disease is no worse than a mild case of the flu. As a result, some believe that the government is using COVID-19 as an underhanded way to reduce personal liberty by eventually making vaccines mandatory. Others argue that the vaccine is ineffective, despite all the evidence to the contrary. Still others charge that the vaccine causes dangerous side effects or has other undesirable characteristics.

Many of the debates over the COVID-19 vaccine reflect discussions around another controversial vaccine. The measles, mumps, and rubella (MMR) vaccine was developed in the late twentieth century to fight three longtime childhood diseases. Like the vaccines for smallpox and COVID-19, the MMR vaccine has been extremely effective. Once, four hundred thousand Americans contracted measles every year, but by the 2010s those numbers had dropped to several hundred cases annually. Beginning in the 1980s, however, there were persistent rumors that administering the MMR vaccine to small children could cause them to develop autism, a disorder that impairs communication and social interaction. Accordingly, a significant number of parents refused to allow their children to receive the vaccine.

Andrew Wakefield

The anti-MMR position received an apparent boost from science in 1998. That year, a British doctor named Andrew Wakefield published a research study in a well-respected medical journal, the *Lancet*. Wakefield and his coauthors studied a group of children with autism and intestinal complaints. The article concluded that most of the children had been diagnosed with autism and stomach issues shortly after receiving the MMR vaccine. Wakefield's research showed what appeared to be a clear link between the MMR vaccine and these two medical conditions.

Unusually for a medical study, Wakefield's research made headlines. "The study stirred fear and controversy among parents and physicians,"[16] reports a health website. Science, it seemed, had validated the anecdotal concerns of anti-vaccine advocates. Based on the findings of the article, alarmed parents decided not to vaccinate their toddlers. They preferred the relatively low risks presented by measles and mumps to what appeared to be the greater danger of autism. Vaccination rates dropped in Britain and elsewhere following the release of the *Lancet* study.

Other researchers quickly set out to replicate Wakefield's results. This is common scientific procedure. The fact that Wakefield and his team found a connection between autism and the MMR vaccine did not prove that one was present. The initial study was simply a starting point. Further investigation was required before science would accept the validity of the *Lancet* article. If Wakefield's research was reproducible, then it would be necessary to take the MMR vaccine off the market until it could be determined what had gone wrong. If it was not reproducible, then the vaccine could be declared safe after all, and its use could be continued.

Andrew Wakefield speaks to the press in London. Wakefield's study linking autism with the MMR vaccine was later shown to contain falsified data; however, as a result of that study, many people still assert that the vaccine causes autism.

The Bone Health Scandal

Yoshihiro Sato was a medical researcher in the field of bone health, a discipline that focuses on ways to treat and prevent fractures of human bones. He was also one of the biggest frauds in scientific history.

Sato, a resident of Japan, turned out several hundred scientific papers in his career. In the early 2000s other researchers began to wonder how Sato managed to find so many volunteers for his studies. Sato first said that he had neglected to list all the hospitals where he had found his subjects but later acknowledged that he had fabricated data in three of his studies and asked that those papers be retracted.

Sato died in 2016, but concerns about the validity of his research continued. As of 2019 investigators had determined that at least sixty of Sato's studies used fabricated data. These studies, too, have been retracted. Many of them, unfortunately, formed the basis of other research that now must be reexamined as well. Given Sato's death, no one will ever know exactly why Sato committed fraud or why he spent so much time creating fictional studies when he had the experience and background necessary to carry out real ones.

Revealing Fraud

As the months went by, though, no research study came close to replicating Wakefield's results. Large-scale studies involving hundreds of children found no link between autism and the MMR vaccine. Scientists and reporters began looking more closely into Wakefield's study. What they found appalled them. Wakefield, it developed, had falsified his data. The medical information he gave for some of the children in the study was completely inaccurate. As just one of the issues, notes a later account of the case, "the Wakefield study reported that patients experienced their first behavioral symptoms within *days* of MMR vaccination, but their medical records documented these as starting some *months* after vaccination."[17]

Wakefield's coauthors distanced themselves from the study. The *Lancet* eventually retracted the results—that is, it acknowledged that the study had been fatally flawed. Wakefield himself refused to admit any wrongdoing, but evidence continued to mount that his data had been faked. It soon developed that Wakefield had been paid almost $675,000 by lawyers hoping to sue vaccine manufacturers—and that Wakefield also hoped to produce and market his own supposedly safe vaccines. In the end Wakefield's license to practice medicine was taken away by British authorities, who cited his fraudulent behavior and his "callous disregard"[18] for the children who took part in his studies.

But even these measures did not stop the widespread belief that autism and the MMR vaccine were linked. Many families have persisted in believing that there is a connection. According to one study, 25 percent of adults today agree with the statement "Some vaccines cause autism in healthy children,"[19] and more than half are unaware that Wakefield's study has been found to be fraudulent. Following the publication of the original article, reports *Time* magazine, "it took nearly two decades for the UK immunization rates to recover."[20] During those twenty years there were over twelve thousand cases of measles, many of them quite serious, and three deaths. And Wakefield still insists that his research is accurate and warns parents against vaccinating their children.

Manipulating Data

Some of the most significant instances of biased science in history are the result of deliberate bias in which scientists alter their data, ignore important counterexamples, or otherwise twist their results to reach a desired conclusion. Unfortunately, the way science operates can make it easy for this type of bias to carry the day. Scientific inquiry generally assumes that researchers are behaving ethically. Because of that, it can be difficult for even seasoned scientists to recognize deception. "Science . . . is intensely sceptical about the possibility of error, but totally trusting about the possibility of fraud,"[21] writes Arnold Relman, a former editor of the *New England Journal of Medicine*. Indeed, despite early evidence that Wakefield's study was seriously flawed, it took twelve years for the *Lancet* to retract the paper.

Perhaps as a result of this mindset, the history of biased research is a long one. One famous example involves the tobacco industry. As early as the 1940s, scientists knew that cigarette smoking was strongly linked to various health issues, notably lung cancer. In 1964 the US surgeon general issued a lengthy report examining the scientific data and concluding that "cigarette smoking contributes strongly to mortality from certain specific diseases and to the overall death rate."[22] Hundreds of thousands of Americans took the warning to heart and stopped smoking. "At the time, I was a first-year medical student," remembers one physician. "Between two-thirds and three-quarters of my fellow students were smokers. By the time

> "Science . . . is intensely sceptical about the possibility of error, but totally trusting about the possibility of fraud."[21]
>
> —Former medical journal editor Arnold Relman

US surgeon general Luther Terry is pictured at a 1964 news conference in Washington, DC, holding a government report that calls smoking a health hazard. Tobacco companies quickly funded studies that cast doubt on the report.

The Mertonian Norms

The ideal behavior for scientists is often expressed in a set of values known as the Mertonian norms, after American social scientist Robert Merton, who developed them in 1942. As described in Stuart Ritchie's book *Science Fictions: How Fraud, Bias, Negligence, and Hype Undermine the Search for Truth*, the Mertonian norms are as follows:

> First, *universalism*: scientific knowledge is scientific knowledge, no matter who comes up with it—so long as their methods for finding that knowledge are sound. The race, sex, age, gender, sexuality, income, social background, nationality, popularity, or any other status of a scientist should have no bearing on how their factual claims are assessed. . . . Second, and relatedly, *disinterestedness*: scientists aren't in it for the money, for political or ideological reasons, or to enhance their own ego or reputation. . . . The third is *communality*: scientists should share knowledge with each other. This principle underlies the whole idea of publishing your results in a journal for others to see. . . . Lastly, there's *organised scepticism*: nothing is sacred, and a scientific claim should never be accepted at face value.

Quoted in Stuart Ritchie, *Science Fictions: How Fraud, Bias, Negligence, and Hype Undermine the Search for Truth*. New York: Metropolitan, 2020.

---

we graduated, only 10% remained smokers. The report was one big reason why."[23]

The science behind the links between smoking and poor health, along with the growing drop in cigarette use, presented a problem for tobacco companies. The report threatened their profits—and their very existence. In response, tobacco companies took several measures, including stepping up advertising in an effort to associate smoking with young, attractive people and a healthy lifestyle. They also did their best to highlight scientific studies that cast doubt on the surgeon general's report. The companies reasoned that if they could publicize studies that reached other conclusions, perhaps smokers would be discouraged from quitting. As a 1969 tobacco company memo expressed it, "Doubt is our product."[24]

Tobacco companies, of course, had every right to promote studies that supported their position. What made their behavior ethically questionable, though, was their funding of studies that fabricated or manipulated data to support their perspective. For example, numerous studies have demonstrated that secondhand smoke causes disease, but the tobacco industry commissioned studies designed to counter this research. The industry's goal for these studies, reports the *European Journal of Public Health*, was to "find ways to conduct animal experiments in such a way that they would not find the known effects of second-hand smoke."[25] The results of the research, then, were known in advance—hardly a hallmark of unbiased scientific investigation.

The tobacco companies have suffered consequences for this behavior. Several medical journals, among them the *European Journal of Public Health* and the *Journal of Health Psychology*, have stopped accepting scientific papers funded even in part by money from the tobacco industry. "It is time to cease supporting the now discredited notion that tobacco industry funded research is just like any other research,"[26] the editorial board of the *British Medical Journal* wrote in 2013. Yet just as with the Wakefield study of vaccines, some members of the general public have been fooled by the scientific veneer of these studies.

> "It is time to cease supporting the now discredited notion that tobacco industry funded research is just like any other research."[26]
>
> —Editorial board of the *British Medical Journal*

Hiding Undesirable Results

Tobacco is not the only example of a business that has used biased science to increase profits or its chances of survival. The fossil fuel industry is another. Even as scientists worldwide warn again and again of the dire threats posed by climate change, companies that deal in coal, oil, and natural gas have worked hard to maintain their share of the energy market. Like tobacco companies, fossil fuel corporations have often failed to act in an unprejudiced manner where scientific studies are concerned.

> "They knew their product was bad, and they were lying to the public."[27]
>
> —Law professor Daniel Farber

"They knew their product was bad," charges law professor Daniel Farber, "and they were lying to the public."[27]

In the case of energy companies, perhaps the most significant issue has been the hiding of studies suggesting that fossil fuels are harmful. As early as 1979 the Exxon Corporation sponsored a study that investigated the consequences of continuing to burn fossil fuels. The study's conclusions were dire. Continuing to rely on oil and similar fuels, the report explained, "will cause dramatic environmental effects." As the report concluded, "The potential problem is great and urgent."[28] Other studies carried out by Exxon and comparable companies came to similar conclusions. By the 1980s, internal corporate documents report, these companies had clear

The burning of fossil fuels is harmful to the environment. There is evidence that fossil fuel companies have known this for many years but kept that information a secret in order to protect their businesses.

evidence that the widespread use of coal, oil, and gas was having a damaging impact on climate.

But no one outside Exxon and its fellow fossil fuel companies learned of these conclusions. Rather than print these studies in reputable scientific journals—or indeed, in any publication at all—the fossil fuel industry hid them. Like the tobacco industry, these companies argued that the science was unsettled and latched onto studies that seemed to prove their point, even when they knew the truth. "Exxon was publicly promoting views that its own scientists knew were wrong," reports former Exxon consultant Martin Hoffert. "They deliberately created doubt when internal research confirmed how serious a threat [climate change] was."[29]

Scientific research begins with observations, which then lead to hypotheses—potential explanations for what is observed. Scientists then design experiments to test their hypotheses. That requires further observation, careful recording of the results, and sometimes a revision of the original hypothesis. Ethical and unbiased science will follow the science wherever it leads, even if a favored hypothesis cannot be confirmed by the data. But in the real world, that is not always the way science works. Scientific research can be subject to cynical manipulation and even outright fraud. Indeed, some of the worst examples of corrupted and biased science are the result of people misusing the scientific model for personal or corporate gain.

CHAPTER THREE

Artificial Intelligence and Algorithms

Human beings are notoriously subjective. It is easy for people to believe they are behaving in an impartial way, only to discover that they are not. Studies show that teachers, for example, often punish Black students more severely than White students for the same offenses. A disruptive White student may get a warning, while an equally disruptive African American student is sent to the principal's office. Similarly, women often receive lower salaries than men even when they are doing the same type of work and have roughly similar backgrounds and experiences.

Disparities such as these spring from what psychologists call implicit bias, or prejudices that lurk below human awareness. "Most people have multiple implicit biases they aren't aware of,"[30] says psychologist Anthony Greenwald. While it is possible for people to make themselves aware of these hidden prejudices, recognizing them takes effort—and combating those prejudices is more difficult still. But although people are far from unbiased, there may be an alternative. What if, instead of trying to fight these implicit biases, we could bypass them altogether? That is the theory behind the use of artificial intelligence—computers and devices able to carry out tasks typically associated with human beings.

Artificial intelligence uses algorithms, or sequences of steps, to make decisions. A computer can determine salaries for men and women based entirely on the worker's experience. Artificial intelligence can mete out punishments for student misbehavior without regard to the student's racial background. Artificial intelligence, in short, should be a scientific and unbiased solution to the problem of human prejudice. Indeed, the past few decades have seen an enormous rise in the use of artificial intelligence. Much of that rise is attributable to the fact that artificial intelligence is quicker and often cheaper than paying a human being to do the same task, but the notion that artificial intelligence is more equitable plays a role as well.

> "Most people have multiple implicit biases they aren't aware of."[30]
>
> —Psychologist Anthony Greenwald

In fact, however, it is not. Artificial intelligence is based on algorithms and data. Those algorithms are created by people who, by virtue of being human, have implicit biases. The data given to these machines and devices also comes from people, who make choices about how to program the machines—based again, in part, on their own implicit biases. Artificial intelligence, then, is only as unbiased—or as biased—as the people who create it. And in many cases, sexism and racism remain easy to see even when the process is taken out of the hands of human beings. Artificial intelligence and the algorithms that power it are not reducing or erasing human biases. Instead, the opposite may be true. "When it comes to algorithm-driven products," writes feminist advocate and author Caroline Criado Perez, "it's making our world even more unequal."[31]

Medical Schools and Risk Assessment

The problem of bias in artificial intelligence is an old one. As early as the 1970s, for example, a British medical college developed a computer program to simplify the job of choosing applicants to interview. The computer could sort through applications in a fraction of the time it would take human admissions officers to do the same task. And ideally it would do so in a way free of the implicit bias common to human beings. That, however, was not the result. In 1986 it was discovered that the program itself was biased against women and against candidates with names that sounded non-European. "As many as 60 applicants each year . . . may have been refused an interview purely because of their sex or racial origin,"[32] concluded an investigation by Britain's Commission for Racial Equality.

It turned out that the algorithm used to choose candidates to interview was designed to match the candidates selected by human admissions officers. These admissions officers showed measurable prejudice against female candidates and against candidates from non-European ethnicities. Those prejudices may well have been unconscious, but they were nonetheless carried over into the algorithms behind the computer program. "Using an algorithm didn't cure biased human decision-making," notes

This picture shows students misbehaving in class. Research shows that even when teachers believe they are treating all their students equally, Black students are often punished more severely than White students for the same offenses.

Job Applications and Artificial Intelligence

Many companies use algorithms in the hiring process. Like using computers to determine who should be interviewed at a medical school, the idea is to save time, along with removing human error and bias from the process. But just as with medical school interviews, artificial intelligence has proved at best a mixed success. Various job application systems have been shown to discriminate on the basis of race, gender, and disability. As academic dean Jelena Kovacevic puts it:

> The bias usually comes from the data. If you don't have a representative data set . . . then of course you're not going to be properly finding and evaluating applicants. . . . For example, if Black people were systematically excluded [in] the past, and if you had no women in the pipeline in the past, and you create an algorithm based on that, there is no way the future will be properly predicted. If you hire only from "Ivy League schools," then you really don't know how an applicant from a lesser-known school will perform, so there are several layers of bias.

Quoted in Sarah K. White, "AI in Hiring Might Do More Harm than Good," CIO, September 17, 2021. www.cio.com.

---

an article in the *Harvard Business Review*. Then again, the article points out, "simply returning to human decision-makers would not solve the problem either."[33]

The goal, then, is to delink artificial intelligence from the biases carried by humans. But even though several decades have passed since the British medical school prejudices were discovered, making algorithms truly objective and unbiased has proved extremely difficult. One example involves risk assessment for criminal behavior. Several corporations, among them a company called Northpointe, have developed computer systems to evaluate how likely a criminal defendant is to continue to commit crimes. Based on answers to questions such as "How often did you get in fights while at school?,"[34] Northpointe grades defendants on a scale of 1 (low risk) to 10 (high risk). Many courts use Northpointe's system to help determine the length of a criminal's sentence.

Unfortunately, risk assessment scores used by Northpointe and other companies are not very accurate. In one Florida study just 20 percent of the defendants predicted to commit more violent crimes actually did so. Also of concern is the fact that these scores show strong racial disparities. African American defendants are much more likely than White defendants to be scored as high risk. That is true even when the White defendant's crimes seem much greater. In one example from the Florida study, a Black female teenager with several juvenile offenses, including stealing a bicycle, was given a rating of 8 on Northpointe's scale, while a White male offender with a history of armed robbery received a score of 3.

No one knows exactly how or why these scores are skewed by race. None of Northpointe's questions address race specifically, and the algorithms that underlie these programs are proprietary. That means they belong to the company that developed them and have not been opened to public scrutiny. But there is no question that programs like Northpointe's exaggerate the risk that Black people will reoffend while underestimating the risk for White people. Among others, former US attorney general Eric Holder and investigative journalism website ProPublica have charged that many defendants have been given longer or shorter sentences than they might have otherwise gotten based entirely on these scores. As Holder and others note, it is a cause for concern when those longer sentences go disproportionately—and erroneously—to Black defendants.

Voice Recognition Systems

Voice recognition software is another form of artificial intelligence that is prone to bias. Among other things, this technology allows people to use their voice in place of passwords to gain access to social media, bank accounts, and electronic devices such as smartphones and tablets. Voice recognition ought to be a significant improvement from password management. Not only does it eliminate the need to recall complex and often arbitrary passcodes—no easy task, as anyone with multiple password-protected accounts knows—but voice recognition is much more secure and much less subject to being hacked.

A girl uses voice recognition technology on her cell phone. Researchers have found that voice recognition software is prone to bias.

However, voice recognition software does not work equally well for everyone. In particular, these systems have trouble interpreting the voices of women. A study from 2016 showed that the most accurate speech recognition system then on the market was 70 percent better able to recognize the speech of men than the speech of women. Anecdotally, too, many women have expressed frustration with their inability to get these systems to respond properly to their vocal commands. When Detroit resident Rebekah Page-Gourley bought a new car, for instance, the voice system could not seem to understand her at all. "I do find myself screaming at it sometimes," she says. "I look like a crazy person screaming at my car."[35] In contrast, the voice system in Page-Gourley's car responded quickly and accurately when her husband spoke to it.

No one knows exactly what the issues are with voice recognition systems, in part because—like risk assessment software—the algorithms they use are proprietary. But evidence suggests that voice recognition software is "trained" primarily using the voices of men. That makes sense. For one thing, recorded voices of men are easier to find; a study of one group of widely available recordings found that over two-thirds featured male voices. In addition, these technologies tend to be developed in settings where men predominate. Whether they are aware of it or not, the men in charge of testing and training voice recognition programs may think of male voices as the default. Their job, then, becomes to make voice recognition systems that can recognize and respond not to human voices in general, but to male voices in particular.

The difficulty voice recognition software has with women's voices is not merely an annoyance. Imagine a woman alone in a car that has broken down on a dark country road late at night, trying to use her voice to open an app on her smartphone. Or suppose that a doctor has used voice software to record information about a patient. Research shows that the notes of women doctors are less likely to be transcribed accurately by these programs in comparison to the notes taken by men. "The fact that men enjoy better performance than women with these technologies means that it's harder for women to do their jobs,"[36] points out language researcher Rachael Tatman. Moreover, medical transcription errors carry potentially negative outcomes for the patient if they are not caught and corrected.

> "The fact that men enjoy better performance than women with these technologies means that it's harder for women to do their jobs."[36]
>
> —Language researcher Rachael Tatman

Facial Recognition Technology

Facial recognition software is another type of artificial intelligence that can replace a passcode. Facial recognition technology, however, is also commonly used for other purposes. Law enforcement agencies use it to scan for wanted criminals in crowds or to help them determine the identities of lawbreakers. Facial recognition

Congress and Flawed Facial Recognition

In 2018 the American Civil Liberties Union (ACLU) performed a test using a widely available facial recognition program called Rekognition. The organization compared photos of the 535 members of Congress with images of 25,000 known criminals taken from a Rekognition database. The test returned twenty-eight positive matches—that is, in twenty-eight cases it identified a member of Congress with a picture in the database. Every one of these was matched in error. The database actually did not include any photos of congressional lawmakers.

Not only were there plenty of errors in the results, the ACLU notes, but there was a distinct racial bias as well. "The false matches were disproportionately of people of color," the ACLU explains, "including six members of the Congressional Black Caucus." In all, 39 percent of the errors involved non-White members of Congress, though they together make up just 20 percent of the total membership. "These results," the ACLU concludes, "demonstrate why Congress should join the ACLU in calling for a moratorium on law enforcement use of face surveillance."

Jacob Snow, "Amazon's Face Recognition Falsely Matched 28 Members of Congress with Mugshots," *Free Future* (blog), American Civil Liberties Union, July 26, 2018. www.aclu.org.

---

software was used to identify those who took part in the January 6, 2021, riot at the US Capitol. Social media sites use facial recognition to enable tagging of people in photographs. Some employers use facial recognition to permit their employees to enter workplaces. Even some colleges are using this technology to track class attendance. "Don't even think of sending your brainy roommate to take your test,"[37] quips a website belonging to computer security company Norton.

At first glance, facial recognition technology seems like a near-perfect system. As with voice recognition, facial recognition software dispenses with the need to memorize and safeguard complicated passwords. And the use of this technology in law enforcement seems like a benefit for all who want a crime-free society. However, facial recognition technology has its drawbacks. For example, its use raises significant privacy concerns. Many people are unwilling to have their faces photographed, scanned, and stored without their permission. In a world filled with facial recognition technology, "what you do and where you go might no longer be private," Norton points out. "It could become impossible to remain anonymous."[38]

Moreover, facial recognition technology is not equally accurate for all groups of people. In one 2018 study, for example, these systems were nearly 100 percent accurate in associating light-skinned men with their photos—but correctly identified dark-skinned women only about two-thirds of the time. The following year another study found that Black and Asian men were 100 times more likely to be misidentified by facial recognition software than were White males. The reason, again, may have to do with the fact that these systems are developed in environments that a 2019 paper by artificial intelligence researchers at New York University describe as "extremely white, affluent, technically oriented, and male."[39] There are other issues too. As journalist Alex Najibi notes, "Default camera settings are often

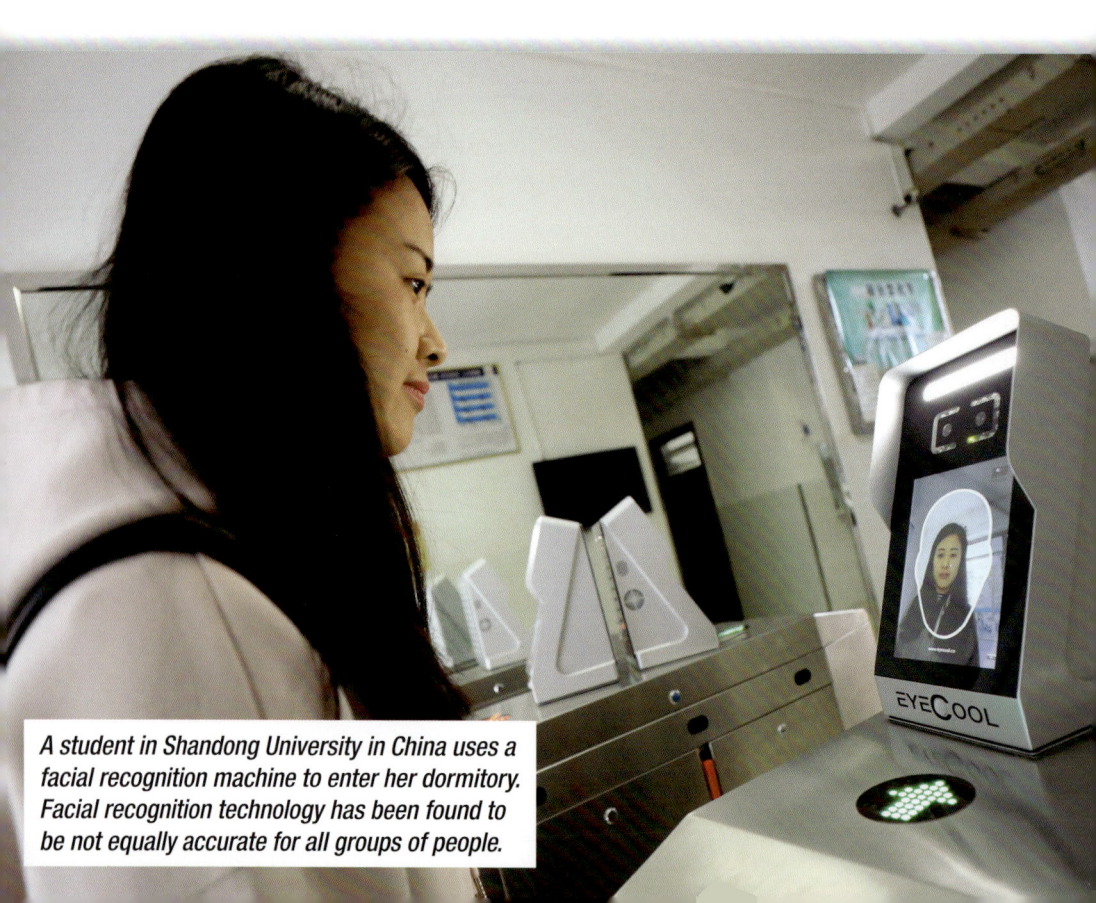

A student in Shandong University in China uses a facial recognition machine to enter her dormitory. Facial recognition technology has been found to be not equally accurate for all groups of people.

not optimized to capture darker skin tones, resulting in lower-quality database images of Black Americans."[40]

At best, these discrepancies are an annoyance for women and people of color who are trying to use facial recognition to get access to their workplace or a social media site. At worst, inaccuracies regarding facial recognition can lead to serious problems. Software, for example, "could misidentify students suspected of fighting, skipping class, or otherwise breaking the rules," writes advocacy organization Human Rights Watch. "The wrong children could be investigated or punished."[41] Perhaps most seriously, there are cases in which men of color have been arrested for crimes they did not commit, simply because their faces were erroneously matched with those of actual criminals.

Given the widespread nature of implicit bias in human beings, artificial intelligence may offer hope of eliminating prejudice in daily life. But it has proved surprisingly difficult to create unbiased algorithms. Whether dealing with law enforcement issues, access to financial accounts, or admission to higher education, it is essential that algorithms and artificial intelligence deal fairly and appropriately with diverse groups of people. That will require new ways of thinking on the part of people who program these devices and develop these technologies. To do otherwise, though, is a missed opportunity—and a misuse of science.

> "Default camera settings are often not optimized to capture darker skin tones, resulting in lower-quality database images of Black Americans."[40]
>
> —Journalist Alex Najibi

CHAPTER FOUR

Sexism and Design

The design of common objects is often based on science. Corporations and governments spend billions of dollars every year trying to ensure that products ranging from cars to can openers are safe, efficient, and easy to use. It is not a coincidence, for instance, that dining room tables in the United States almost always have a height of 28 to 32 inches (71 to 81 cm). Rather, they are this tall because research shows that most Americans are comfortable sitting at a table this size. In another example, the accelerator and brake pedal are arranged the way they are on a typical vehicle because research has shown that this configuration makes shifting from one to the other relatively easy for drivers. Even the designs of vegetable peelers and other kitchen tools are a function of research into shapes and sizes that fit easily in the user's hand. The list goes on.

Most of these products are designed with a particular person in mind—a sort of "standard American" who will have no difficulty using the product as it was intended. On the surface, basing product design on a standard American seems both scientific and sensible. But in fact, where design is concerned, the "average" or "standard" American is likely to be a good deal less representative of the population than people typically think. The reason is that products in the United States and beyond are usually designed for men. All too often, the needs and bodies of women are not considered when products are being designed. Science is supposed to be unbiased and neutral, but in the case of design, it is anything but. Where design is concerned, says Caroline Criado Perez, "the average woman is an outlier."[42]

Treadmills, Cell Phones, and Space Suits

Design features that fail to take women into account can often be annoying. A good example is the treadmill. On one level, treadmills seem to be perfectly designed for almost anyone. After all, a treadmill can be set to move at different speeds and at different inclines to meet the needs of a largely sedentary person or a professional athlete. Most treadmills, too, have displays that show how many calories users are burning during their workouts, and this display is adjustable. Heavier people burn more calories than lighter people given equal exercise, so by keying in their weight, users can ensure that the machine is tracking the correct number of calories they are burning.

There are two problems with treadmills and calories, though, one universal and the other specific to women. The universal issue is that people in general vary considerably in the number of calories they burn through exercise. As Criado Perez puts it, "The calorie count on treadmills is perfect for practically no one."[43] But incorrect calorie counts are a specific issue for women. Regardless of size, men and women do not burn calories at the same rate given the same amount of exercise. The calorie counter on a treadmill shows the number of calories a typical man of a given weight would be burning—not the number of calories for a typical woman. As a result, women generally burn fewer calories than the machine says they do.

Sexism in design also affects cell phones. The average cell phone of the 2020s is an appropriate size to fit comfortably in the hand—of a man, but not a woman. The average man has a hand about 1 inch (2.5 cm) larger than the average woman's. That

> "The average woman is an outlier."[42]
>
> —Author and feminist advocate Carline Criado Perez

makes it difficult for women to do certain things with cell phones, like take photographs with one hand—an activity that is relatively easy for a man. "Companies have got to get better at recognizing that their idea of normal should account for all their customers,"[44] argues a British member of Parliament. Another customer, complaining directly to the Apple corporation, writes succinctly, "I can't grow my hands any bigger. Please make a smaller [iPhone]."[45]

Another example of sexism in design involves space suits, which are used in space walks at the International Space Station. "What does it take to don a spacesuit and venture out on such a technical and dangerous mission?" asks science professor Steven Moore. "Surprisingly, one of the main criteria . . . is body size."[46] The National Aeronautics and Space Administration used to make space suits in five sizes. In the early 1990s, though, the agency stopped manufacturing the two smallest sizes. About one-third of the women astronauts of the time could only fit into these sizes. The result was a drastic reduction in the number of female astronauts who could go on a space walk, for which a well-fitting space suit is essential.

Trucks and many other products in the United States and other countries are often designed for the needs and bodies of men, not women.

Anne McClain's Space Walk

In March 2019 the National Aeronautics and Space Administration (NASA) planned a space walk at the International Space Station featuring two American astronauts, Anne McClain and Christina Koch. The space walk was expected to be historic: though more than a dozen female astronauts had gone on space walks before, there had never been a space walk in which both astronauts were women. But the space walk had to be canceled when McClain discovered that she could not fit properly into a large size space suit, as she had believed she could. While there were two medium space suits on board the space station, only one was configured for space walking at the time. Thus, a male astronaut took McClain's large suit, and the all-female space walk did not take place as scheduled.

Though the final decision was McClain's, many observers blamed NASA for not ensuring that there were two properly configured medium suits aboard the station. "They're doing their part to do their jobs," says former NASA medical adviser Saralyn Mark of women astronauts. "They're putting in the effort, they're putting in the physical exhaustion, they're putting in the mental fortitude. Then give them the equipment so they can do their jobs."

Quoted in Emily Dreyfuss, "The Failure of NASA's Spacewalk Snafu? How Predictable It Was," *Wired*, March 28, 2019. www.wired.com.

Military Equipment

Sexism in the design of treadmills and cell phones are certainly an irritation but arguably does not have serious potential consequences for health and safety. That is not true, however, for all issues of design. Military equipment and uniforms, for instance, are not well suited for women's bodies—and yet are of the utmost importance in keeping soldiers and other military personnel safe. Until 2016 most military uniforms came only in sizes too big for many women serving in the armed forces. Even when the US Army introduced smaller uniform sizes in 2016, it was still difficult for women to find appropriately sized boots and helmets.

> "We've . . . heard testimony from servicemembers who remove their body armor in combat because it is too heavy."[47]
>
> —US representative Niki Tsongas

Body armor has been a particular issue for many women in combat. US representative Niki Tsongas, for example, has pushed the military to develop lightweight body armor suitable for women's bodies. "We've . . . heard testimony from servicemembers who remove their body armor in combat because it is too heavy,"[47] Tsongas reports. Moreover, Tsongas notes, some women have found that the weight of their body armor prevents them from lifting their arms high enough to fire their weapons properly. Both are problems that should give everyone pause. "There are real risks to not moving ahead [to fix these issues] in a very expeditious way,"[48] Tsongas says.

US Air Force captain Rebecca Lipe had equally significant concerns regarding armor that did not properly fit her body. Lipe was stationed in a combat zone in the Middle East for a year and spent much of that time wearing body armor. Unfortunately,

Women in the military often complain that their body armor does not fit them well. In this 2021 picture, Senior Airmen Rosealine McDonald and Deanna Mayhew show their new body armor, which is designed to better fit a female body.

the armor never fit correctly. She ultimately had to place foam rubber under the shoulder straps to keep the armor in place, and even then it did not completely protect all her vital organs. "Theoretically if you get shot in the side, there's side panels," Lipe explains, "but I couldn't even wear those, because I couldn't get [them] to fit or stay on correctly."[49]

Women in Science

Health and safety can also be an issue with regard to scientific gear and equipment. Many women in science have complained that the clothing and protective gear available to them is designed with the male body in mind. Jessica Mounts, a biologist in Kansas, has been particularly vocal in her displeasure about the design of clothing available to people in her profession. "Most of the equipment I've used has been designed for men," says Mounts. "The problems caused aren't simply an annoyance—they all go back to personal safety." As she points out, "Clothing that is too loose gets caught in moving equipment. Boots that are too big mean tripping and falling."[50]

> "Clothing that is too loose gets caught in moving equipment. Boots that are too big mean tripping and falling."[50]
>
> —Biologist Jessica Mounts

Mounts is not alone in her frustration—and her concerns for her health and safety. When she posted her concerns on social media, she heard from dozens of other women with similar complaints. One of her respondents, for example, reported having to use lab goggles that did not fit her face properly. Another woman complained that the only surgical gowns available to her were so large they had to be wrapped around her more than once, causing overheating. Similarly, a biologist was unable to find properly sized steel-toe boots for the work she was doing. In some cases employers may not keep sizes on hand that fit women. In other cases gear may be completely unobtainable because it is simply not produced in appropriate sizes for women.

The situation has improved somewhat. Women scientists can find some needed apparel in sizes that fit them. For example, Mounts cites wading boots and personal flotation devices sized for women, once unavailable but both manufactured today. "Things have improved marginally," Mounts notes. "That said, we have a long way to go." The problem, she points out, goes beyond putting women scientists in danger. Rather, the underlying message is to discourage women from becoming scientists altogether. "When only men are allowed to work [as scientists]," Mounts says, "the only equipment available is made for them. When women want to work, the equipment isn't there and is a tangible symbol of the greater issue of systemic sexism in a society designed, in general, by and for men."[51]

Automotive Safety

In the cases of science and the military, there is a clear reason why design has been focused on the needs of men over the years. Until relatively recently, there have not been many women in science or in the military. That does not excuse the lack of appropriate lab coats or body armor for women, but it does begin to explain why women in these fields have struggled to find appropriate gear. There is no such rationalization, though, for other design issues, such as those involving automotive safety. Though women have been driving and riding as passengers almost since cars were invented, studies of safety equipment have focused almost entirely on male bodies—with predictable consequences for women involved in vehicular accidents.

Through the first few decades of the twentieth century, very little attention was paid to safety. Driving could be extremely dangerous, and crashes were common. Early efforts to reduce accidents tended to focus on driver behavior and highway design, but as time went on, traffic experts began to realize that automotive design also played an important role in the number of crashes. Too many cars during the first half of the twentieth century had features that proved hazardous to drivers and passengers alike. These included rigid steering wheels that did not collapse on impact, dashboards without padding, and glass that shattered into sharp pieces when broken. By the 1950s engineers were increasingly aware of these issues and the role they played in causing accidents and making those accidents more serioius than they needed to be.

One important development in auto safety took place in the 1960s, when automotive engineers began using crash test dummies to model what happens to drivers and passengers in vehicular accidents. Crash test dummies are strapped into cars that are then subjected to collisions of various types. What happens to the dummies gives engineers a good idea of what might happen to a human occupant of the vehicle in that type of crash— and highlights which design features need to be improved. Crash

This picture shows a crash test dummy after a simulated car accident. For years, engineers used just one crash-test dummy, which was based on the size of the average male.

CPR Mannequins

In 2017 the Perelman School of Medicine at the University of Pennsylvania ran a study investigating what happened to people who had heart attacks in public places. The study concluded that men were 23 percent more likely than women to survive these heart attacks. Researchers believe that one factor is the unwillingness of strangers to perform cardiopulmonary resuscitation (CPR) on women. The thinking is that people who know CPR are reluctant to touch the breasts of women they do not know—a requirement for using CPR on a female stranger.

A new device called the Womanikin aims to change this situation. The Womanikin, its developers explain, is "a universal attachment that can easily be slipped over the common flat-chested CPR mannikin." The idea is to give CPR learners experience in working on women's bodies so they are not put off by the possibility of touching a breast while performing the life-saving procedure on a woman. As the manufacturer puts it, "For years, our society has made touching breasts taboo. And while CPR training has become more commonplace, we still learn on male torsos. Our goal is to bridge the gap in CPR training by normalizing a woman's figure."

Womanikin, "Learn on a Woman, Save a Woman." https://womanikin.org/About.

---

test dummies are effective. According to one study, the changes made to cars as a result of crash tests involving dummies have saved well over three hundred thousand lives.

Disparities

For years, however, engineers used just one crash test dummy—and it was based on the size of the average male. The average man is significantly taller and heavier than the average woman. As Criado Perez puts it, "The dummy also has male muscle-mass proportions and a male spinal column."[52] What happens to a dummy in a crash thus reflects the experience of an average male involved in that accident. But it does not reflect the experience of most women. Automotive safety decisions are being made based on what keeps men safe—not on what keeps women safe.

And indeed, there are significant differences between men and women where crash-related injuries are concerned. According to studies done over the past two decades, women are nearly 50 percent more likely than men to be seriously hurt in a car accident. They are also 17 percent more likely to be killed. The disparity has been known for years, but little has been done to address it. Though female dummies are now available, they tend to be smaller versions of the male dummies, which still does not address the different shapes and anatomies of women. Like the issues regarding scientific gear and military equipment, the use of crash test dummies that model male bodies is not just an annoyance—it is actively harming and even killing women.

The science of design is a natural consequence of male dominance in technical fields. But as Criado Perez points out, it also is an indictment of science as it is currently practiced. "Our society positions science as neutral," she says, "as objective and free of bias. Science deals in facts. In truth. Only, now it turn[s] out that our cultural positioning of men as the default humans [is] corrupting science."[53] In this view, accounting for women's bodies and needs in design will not only improve women's lives, it will improve science itself.

CHAPTER FIVE

Medicine and Medical Trials

Medical ethics have come a long way since the days of Henrietta Lacks and the Tuskegee syphilis study. Sparked in large part by the news of the Tuskegee experiment, Congress passed laws making informed consent an absolute requirement for carrying out most medical studies. Universities and hospitals have established their own institutional review boards, which examine all proposed studies closely to make sure no one's rights are being violated. "The rules and policies for human subjects research have been reviewed and revised many times since they were first approved," writes the CDC, "and efforts to promote the highest ethical standards in research are ongoing."[54]

Adoption and enforcement of ethical standards in medical research are important steps in efforts to eliminate bias from this scientific field. This goal, however, has not yet been achieved. Bias still exists in clinical trials for new drugs and in criteria for diagnosing common conditions such as heart attacks and autism, among other areas of medicine.

Clinical Trials

The process of developing new medications and medical devices is long and complex. Medicines and medical devices need to undergo all sorts of tests in laboratories before they can be marketed to the general public. Once drugs and medical devices are deemed ready to be tested on human subjects, they are typically first tested on a few dozen relatively healthy people to ensure that they have no dangerous side effects. This is called a Phase 1 clinical trial. If the intervention appears safe, it advances to a Phase 2 trial. In this step it is used on several hundred people, all of whom have the disease or condition that the intervention is intended to address. Here the emphasis is on effectiveness. Researchers try to determine whether the intervention is helpful in curing or alleviating the disease.

> "The rules and policies for human subjects research have been reviewed and revised many times since they were first approved, and efforts to promote the highest ethical standards in research are ongoing."[54]
>
> —Centers for Disease Control and Prevention

Phase 2 trials can last for several years. They are followed by Phase 3, in which the number of research subjects may be in the thousands and both the safety and effectiveness of the intervention are double- and triple-checked. Only if the intervention passes Phase 3 will it be approved by the US Food and Drug Administration (FDA). Even then, a drug or device may still be monitored in some patients for several years, in case there are side effects that only appear with time. This final monitoring is Phase 4, the last stage of the clinical trial process. It is common for drugs, devices, and other interventions to take ten years or more to pass successfully through all four of the phases of the clinical trial process.

Clinical studies rely on volunteers—people who will agree to take the medication in question during one of these phases and allow their responses to be studied by researchers. Volunteers, however, can be difficult to find. By one estimate, as many as 80 percent of studies have to be delayed for months or even years because of a lack of volunteers. "Most studies do not finish on time due to low patient participation and slow recruitment,"[55] says Laurie M. Ryan of the National Institutes of Health. No more than 4 percent of Americans have ever participated in a medical research study.

A Lack of Diversity

Moreover, those who do take part in a clinical trial are seldom reflective of the diversity of America. Women, for example, make up relatively small numbers in most medical research studies. In a group of thirty-one studies related to congestive heart failure, for example, just a quarter of the subjects were female. The numbers are even lower in studies for certain other conditions. Over a period of several years, for example, women made up only 11 percent of the subjects in trials designed to find a cure for HIV. The numbers are particularly low for Phase 1 and Phase 2 trials; a 2018 scientific paper found that 22 percent of Phase 1 participants overall are women.

These low numbers present some serious problems. Women and men may process drugs differently, due to both body size and body chemistry. Testing mainly men runs the risk of producing medications and other interventions that work perfectly well for most men—but are ineffective or even dangerous for women. Coronary stents, for example, can help patients with cardiac problems by artificially expanding the size of the body's blood vessels.

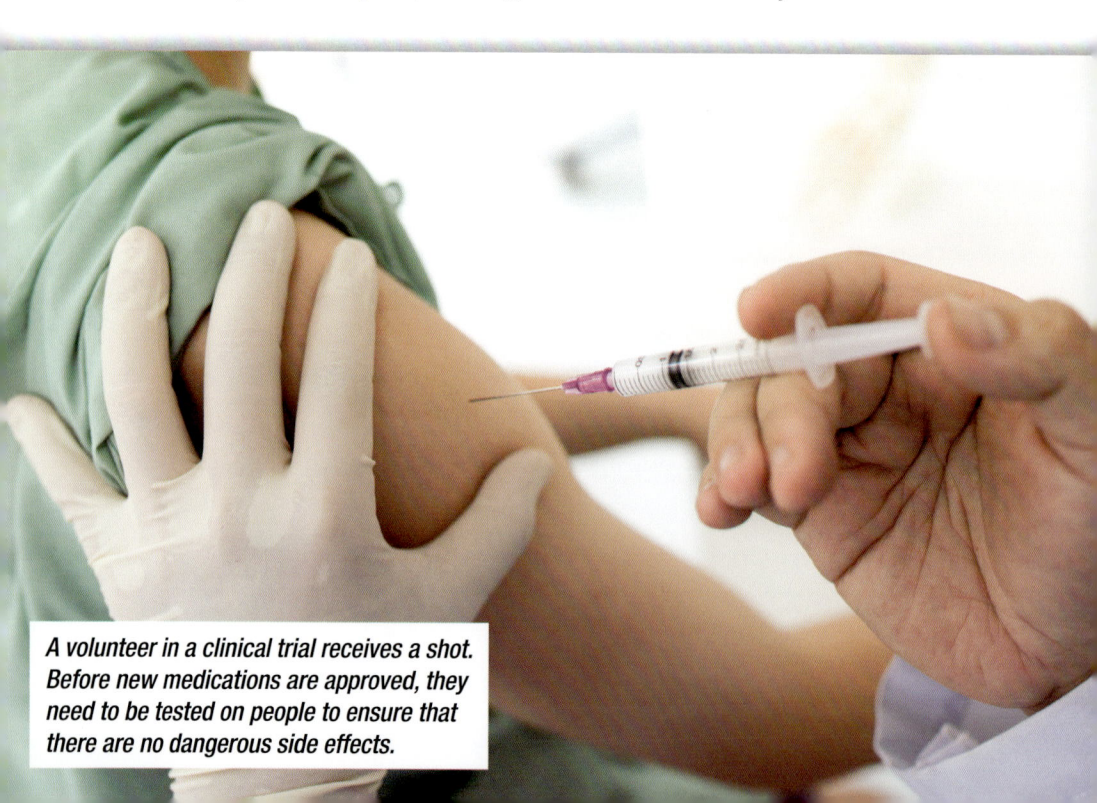

A volunteer in a clinical trial receives a shot. Before new medications are approved, they need to be tested on people to ensure that there are no dangerous side effects.

Women and Clinical Trials

There are several reasons why clinical medical trials tend to have few women. One is simply that taking part in a trial requires time and money, and women are typically both busier and poorer than men. Most trials pay subjects nothing more than transportation costs, if that, so it may be relatively more difficult for women to justify taking time off work to participate in a study. That becomes doubly problematic since many women have substantial family responsibilities as well. It may be a deal breaker if signing up for a study requires paying for care for a child or an elderly relative.

There is a historical reason for excluding women as well. In the early 1960s a drug called thalidomide was briefly approved for use in Europe to treat morning sickness, a side effect of early pregnancy. Tragically, thalidomide was later found to cause severe birth defects in children born to women who took the drug. Using what may have been an overabundance of caution, the US Food and Drug Administration later issued guidelines suggesting that women of childbearing age be excluded from medical research. Though the guidelines are no longer in effect, their impact lingers.

Some studies, though, have found that stents work better in men than in women. Several experts believe that women's issues with stents are directly related to the fact that just 32 percent of the subjects in the original trials were women. Since stents were tested largely on men, this reasoning goes, they were tailored to meet the needs of male bodies rather than those of females.

Ambien, a sleep-inducing medication, is another example. Based on studies with more male representation than female, Ambien was approved at a single dosage for all adult patients. Later, however, researchers discovered that men and women tended to react very differently the morning after using the drug. Men tended to be alert and energetic the next day. Women, on the other hand, were more tired than normal the following morning. "The standard dose produced much higher blood concentrations and longer drug elimination times in women than in men,"[56] says scientist Irving Zucker. The problem was resolved in 2013 when the FDA established a new recommended dose for women that was half the original dosage. But the problem might have been resolved much sooner with a more equitable distribution of test subjects.

Another issue with male-dominated studies involves medicines that might prove useful to women—but are never put on the market because they had no effect on the male subjects on whom the drugs were tested. As Criado Perez puts it, "How many drugs that *would* work for women are we ruling out at phase one trials just because they don't work in men?"[57] No one knows the answer, but it is easy to imagine that hundreds of effective medications were abandoned early in the testing process, not because there was something wrong with the drugs but because there was something wrong with the researchers' pool of subjects.

Like women, men of color are seriously underrepresented in clinical trials. In 2015, for example, a group of researchers discovered that fewer than 2 percent of studies on respiratory illnesses included any non-White people. Overall, only 10 to 20 percent of American clinical trial subjects are people of color, despite the fact that racial minorities make up over 40 percent of the US population. "That leads to a real problem," notes an editorial in *Scientific American*. "The symptoms of conditions such as cancer, heart disease and diabetes, as well as the contributing factors, vary across lines of ethnicity. . . . If diverse groups aren't part of these studies, we can't be sure whether the treatment will work in all populations."[58]

> "If diverse groups aren't part of these studies, we can't be sure whether the treatment will work in all populations."[58]
>
> —Editorial board of *Scientific American*

Making Diagnoses

Just as the medical field fails women and racial minorities by traditionally leaving them out of research studies, medicine often fails the same people where diagnoses are concerned. Part of the problem, again, is that much of medicine's conventional wisdom is based on the progress of disease in men or in White people, rather than in the case of everyone. A good example involves heart attacks. Many people have been taught that heart at-

This picture shows a bottle of Ambien pills. Ambien was initially approved at a single dosage for all adult patients; however, that dosage was based on male-dominant studies and was later found to be too high for women.

tacks usually involve chest pains and left arm numbness. In fact, that is true for men. But it is not accurate for women, especially younger women, for whom heart attacks are more often marked by nausea, breathlessness, and stomach pain. It should come as no surprise, then, that heart attacks in women are 50 percent more likely to be misdiagnosed than heart attacks in men.

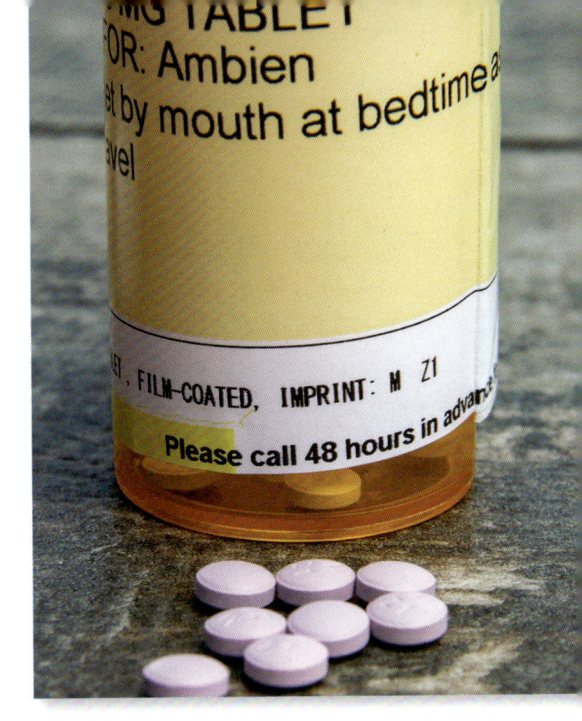

Heart problems are not the only example of misdiagnoses, however. Conditions that affect the brain, such as autism and attention deficit disorder, are likewise underdiagnosed where girls and women are concerned. It was thought at one point that autism was four times more common in boys than in girls. But more recent research suggests that the true numbers are much closer to even. According to Sarah Wild, a British educator who runs a school for girls with autism, "The diagnostic checklists and tests [for autism] have been developed for boys and men, while girls and women present completely differently."[59] Similarly, about 75 percent of girls with attention deficit disorder are believed to be undiagnosed. Girls and women with these conditions often do not get the help they need.

Nor are women the only victims of medical misdiagnosis. Certain diseases, among them diabetes, heart problems, and cancer, often go undiagnosed among Americans of color. Thomas Usher, an African American martial arts instructor from Lexington, Kentucky, suffered from breathing difficulties throughout his childhood, but he was not given the diagnosis of asthma until he was

Medical Algorithms

Algorithms for use in medical settings have become increasingly common in recent years. A computer can use information about a patient's symptoms to determine an appropriate course of treatment. Unfortunately, like much else in medicine and artificial intelligence, these algorithms are biased against people of color, particularly African Americans. The computers that apply these algorithms frequently recommend one course of treatment for Black people and a faster, more aggressive one for White people, even if the underlying symptoms are the same.

That is because one of the data points used by the algorithms involves how much money has been spent on the patient's health care during the previous year. The idea is that more money will have been spent on sicker patients, and sicker patients are more in need of aggressive treatment. Though this may hold true in some cases, in reality White people are more likely to see a doctor when they are sick. Because Black people wait longer to see doctors, their health care spending is lower, and the algorithms believe them to be healthier than they actually are.

---

an adult. "You should have been hospitalized all your life,"[60] he remembers a doctor telling him when he was twenty-six. Today outcomes for many types of cancer are much more favorable for White people than for Black Americans, in large part due to the delay in making a proper diagnosis. Black women are 40 percent more likely to die of breast cancer than White women, for example, and Black men are more than twice as likely to die of prostate cancer than their White counterparts.

Systemic Racism

In the case of racial minorities, at least some of the disparity in diagnoses and outcomes seems to be attributable to systemic racism. The story of Gary Fowler of Detroit is depressingly familiar to many African Americans, especially those who are poor. Fowler contracted COVID-19 early in the pandemic, but despite symptoms that included a high fever and a persistent cough, he could not get a hospital to admit him. He went home, where he died a few days later. "He was begging for his life, and medical professionals did nothing for him,"[61] says Keith Gambrell, Fowler's stepson. To make matters worse, Gambrell's mother soon began suffering from COVID-19-like symptoms, but when Gambrell took her to the hospital, she was not seen until after a White woman who appeared to be much less sick.

Indeed, there is statistical evidence that medical science is often biased against racial minorities, particularly African Americans. One recent study showed that when Black people go to emergency rooms, their wait to be seen by medical professionals is typically more than 25 percent longer than for White people. While some of the disparity may have to do with the numbers of people waiting to be seen, the study's authors believe that racism is a primary cause of the difference. Whether intentionally or otherwise, emergency room personnel are not taking Black people's complaints as seriously as they should.

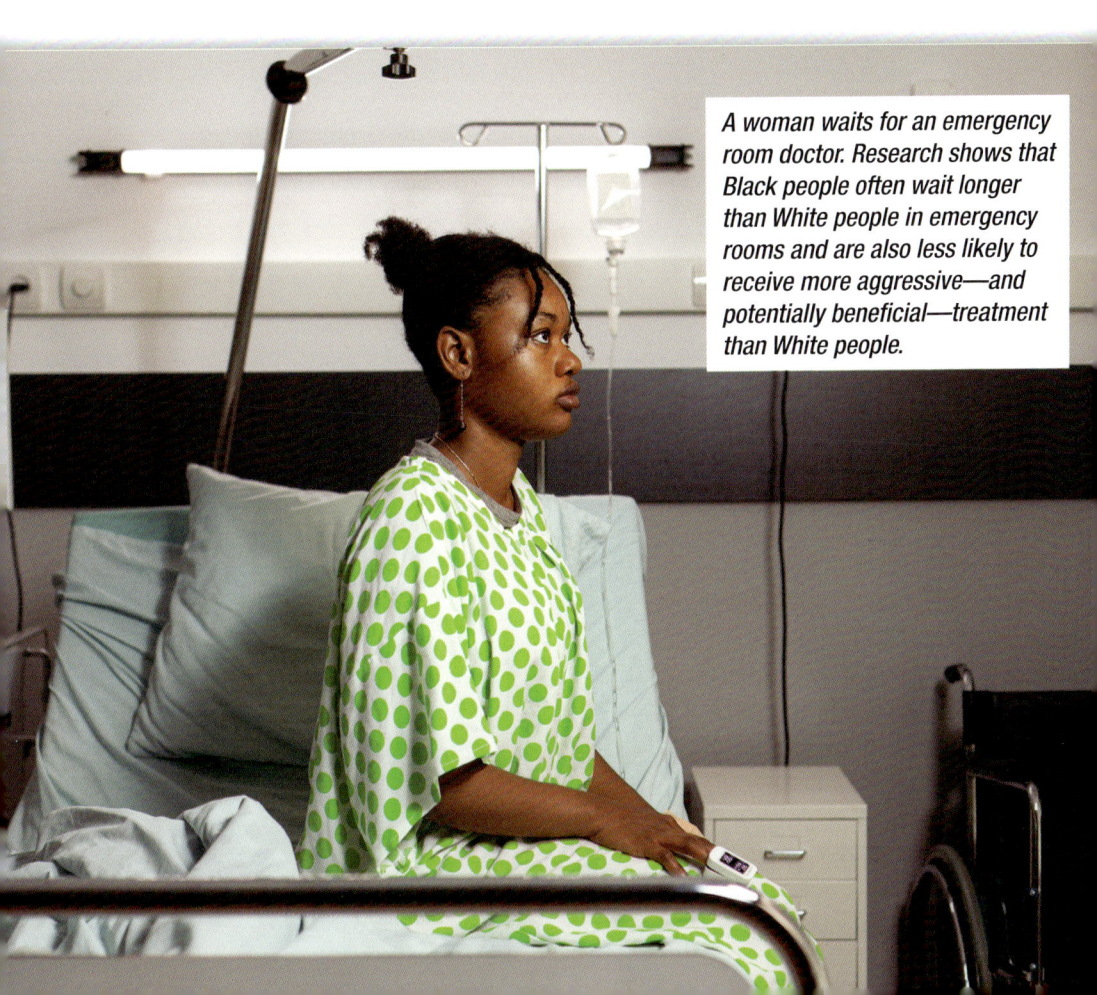

A woman waits for an emergency room doctor. Research shows that Black people often wait longer than White people in emergency rooms and are also less likely to receive more aggressive—and potentially beneficial—treatment than White people.

Alexander Green, a doctor at Massachusetts General Hospital, ran a different study in 2007 to determine the degree of racial bias present in emergency rooms. He gave emergency room doctors information about a patient suffering from chest pains and asked them to recommend an appropriate treatment. Green also included photos of the supposed patient. Some of the doctors saw a picture of a White person, others an image of a Black person. The doctors who thought the patient was White recommended more aggressive treatments—treatments more likely to result in positive outcomes—than those who believed they were seeing a Black patient. "It could be bias, conscious or unconscious, on the part of providers or other staff that work at the site where they're receiving care,"[62] Green says.

Whether excluding women from clinical studies or offering substandard care to African Americans, and whether misdiagnosing common health issues in women or prioritizing the medical care of White people, the science of medicine shows strong racial and gender biases. Americans of all races and genders have the right to expect medical care to be fair and impartial, not to favor one race or gender over another. But that is not currently the case. Rather, medical science is far more sensitive to the needs of White people and White men in particular than it is to the needs of women and racial minorities. As Criado Perez writes in her book *Invisible Women*, "We need a revolution in the research and practice of medicine, and we need it yesterday."[63]

> "We need a revolution in the research and practice of medicine, and we need it yesterday."[63]
>
> —Author and feminist advocate Caroline Criado Perez

SOURCE NOTES

Introduction: Scientific Bias
1. Quoted in Peter Galison et al., eds., *Science in Culture*. New York: Routledge, 2017, pp. 161–62.
2. Quoted in John Horgan, "Darwin Was Sexist, and So Are Many Modern Scientists," *Cross-Check* (blog), *Scientific American*, December 18, 2017. https://blogs.scientificamerican.com.

Chapter One: The Tuskegee Experiment and Henrietta Lacks
3. Quoted in James H. Jones, *Bad Blood*. New York: Free Press, 1981, p. 7.
4. Quoted in Harriet A. Washington, *Medical Apartheid*. New York: Doubleday, 2006, p. 163.
5. Quoted in Jones, *Bad Blood*, pp. 5–6.
6. Quoted in Washington, *Medical Apartheid*, p. 166.
7. Quoted in Jones, *Bad Blood*, p. 10.
8. Quoted in Tuskegee University, "About the USPHS Syphilis Study," 2022. www.tuskegee.edu.
9. Bill Clinton, "Apology for Study Done in Tuskegee," White House, May 16, 1997. https://clintonwhitehouse4.archives.gov.
10. Quoted in Rebecca Skloot, *The Immortal Life of Henrietta Lacks*. New York: Crown, 2010, p. 43.
11. Quoted in Skloot, *The Immortal Life of Henrietta Lacks*, p. 31.
12. Quoted in Skloot, *The Immortal Life of Henrietta Lacks*, p. 41.
13. American Medical Association, "Informed Consent," Nevada Legislature, 2013. www.leg.state.nv.us.
14. Ta-Nehisi Coates, "Henrietta Lacks and Race," *The Atlantic*, February 3, 2010. www.theatlantic.com.
15. Cary Funk et al., "Black Americans Have Less Confidence in Science to Act in the Public Interest," Pew Research Center, August 28, 2020. www.pewresearch.org.

Chapter Two: Scientific Fraud
16. Healio, "Wakefield Study Linking MMR Vaccine, Autism Uncovered as Complete Fraud," February 1, 2011. www.healio.com.
17. Quoted in Public Health Seattle & King County, "MMR, Autism, and Wakefield—Responding to Vaccine-Hesitant Parents," *Issue Brief*, February 2011. https://kingcounty.gov.
18. Quoted in Jonathan D. Quick and Heidi Larson, "The Vaccine-Autism Myth Started 20 Years Ago. Here's Why It Still Endures Today," *Time*, February 28, 2018. https://time.com.

19. Quoted in Public Health Seattle & King County, "MMR, Autism, and Wakefield—Responding to Vaccine-Hesitant Parents."
20. Quick and Larson, "The Vaccine-Autism Myth Started 20 Years Ago."
21. Quoted in Fiona Godlee, "Wakefield's Article Linking MMR Vaccine and Autism Was Fraudulent," *The BMJ*, January 6, 2011. www.bmj.com.
22. Quoted in Kayla Ruble, "Read the Surgeon General's 1964 Report on Smoking and Health," *PBS NewsHour*, January 12, 2014. www.pbs.org.
23. *Harvard Health Blog*, "Surgeon General's 1964 Report: Making Smoking History," Harvard Medical School, January 10, 2014. www.health.harvard.edu.
24. Quoted in Rahul Kanakia, "Tobacco Companies Obstructed Science, History Professor Says," Stanford University, February 13, 2007. https://news.stanford.edu.
25. Quoted in Martin McKee, "Why the *European Journal of Public Health* Will No Longer Publish Tobacco Industry–Supported Research," ResearchGate, 2014. www.researchgate.net.
26. Quoted in McKee, "Why the *European Journal of Public Health* Will No Longer Publish Tobacco Industry–Supported Research."
27. Quoted in Chris McGreal, "Big Oil and Gas Kept a Dirty Secret for Decades. Now They May Pay the Price," *The Guardian* (Manchester, UK), June 30, 2021. www.theguardian.com.
28. Quoted in McGreal, "Big Oil and Gas Kept a Dirty Secret for Decades."
29. Quoted in McGreal, "Big Oil and Gas Kept a Dirty Secret for Decades."

Chapter Three: Artificial Intelligence and Algorithms

30. Quoted in Betsy Mason, "Making People Aware of Their Implicit Biases Doesn't Usually Change Minds. But Here's What Does Work," *PBS NewsHour*, June 10, 2020. www.pbs.org.
31. Caroline Criado Perez, *Invisible Women*. New York: Abrams, 2019, pp. 167–68.
32. Stella Lowry and Gordon Macpherson, "A Blot on the Profession," *British Medical Journal*, March 5, 1988. www.ncbi.nlm.nih.gov.
33. James Manyika et al., "What Do We Do About the Biases in AI?," *Harvard Business Review*, October 25, 2019. https://hbr.org.
34. Quoted in Julia Angwin et al., "Machine Bias," ProPublica, May 23, 2016. www.propublica.org.
35. Quoted in Sharon Silke Carty, "Many Cars Tone Deaf to Women's Voices," *Autoblog*, May 31, 2011. www.autoblog.com.
36. Quoted in Criado Perez, *Invisible Women*, p. 163.
37. Steve Symanovich, "What Is Facial Recognition? How Facial Recognition Works," Norton, August 20, 2021. https://us.norton.com.
38. Symanovich, "What Is Facial Recognition?"
39. Quoted in Johana Bhuiyan, "Facial Recognition May Help Find Capitol Rioters—but It Could Harm Many Others, Experts Say," *Los Angeles Times*, February 4, 2021. www.latimes.com.
40. Alex Najibi, "Racial Discrimination in Face Recognition Technology," *Science in the News* (blog), Harvard University, October 24, 2020. https://sitn.hms.harvard.edu.
41. Human Rights Watch, "Facial Recognition Technology in U.S. Schools Threatens Rights," June 21, 2019. www.hrw.org.

Chapter Four: Sexism and Design

42. Quoted in Ritu Prasad, "Eight Ways the World Is Not Designed for Women," BBC, June 5, 2019. www.bbc.com.
43. Criado Perez, *Invisible Women*, p. 177.
44. Quoted in Christian Gollayan, "Apple's Bigger-Screen iPhones Are Slammed for Being 'Sexist,'" *New York Post*, September 14, 2018. https://nypost.com.
45. Quoted in Gollayan, "Apple's Bigger-Screen iPhones Are Slammed for Being 'Sexist.'"
46. Steven Moore, "NASA's Few Remaining Spacesuits Are Old, and They're Not a Great Fit for Women," *Washington Post*, July 11, 2021. www.washingtonpost.com.
47. Quoted in Nashoba Publishing, "Tsongas Legislation to Improve Body Armor and Prevent Sexual Assault in the Military Passed by the House," *Nashoba Valley Voice* (Ayer, MA), June 4, 2010. www.nashobavalleyvoice.com.
48. Quoted in Travis J. Tritten, "Form-Fitted Body Armor Rolling Out as Combat Roles Expand for Women," *Stars and Stripes*, March 4, 2016. www.stripes.com.
49. Quoted in Ema O'Connor and Vera Bergengruen, "Military Doctors Told Them It Was Just 'Female Problems.' Weeks Later, They Were in the Hospital," BuzzFeed News, March 8, 2019. www.buzzfeednews.com.
50. Quoted in Chris Bell, "One Small Step for Man, but Women Still Have to Leap," BBC, March 28, 2019. www.bbc.com.
51. Quoted in Bell, "One Small Step for Man, but Women Still Have to Leap."
52. Criado Perez, *Invisible Women*, p. 186.
53. Caroline Criado Perez, "The Dangers of Gender Bias in Design," Evoke, July 11, 2019. www.evoke.org.

Chapter Five: Medicine and Medical Trials

54. Centers for Disease Control and Prevention, "Research Implications," April 22, 2021. www.cdc.gov.
55. Quoted in Elly Earls, "Clinical Trial Delays: America's Patient Recruitment Dilemma," Clinical Trials Arena, July 18, 2012. www.clinicaltrialsarena.com.
56. Quoted in Elizabeth Pratt, "We Don't Have Enough Women in Clinical Trials—Why That's a Problem," Healthline, October 25, 2020. www.healthline.com.
57. Criado Perez, *Invisible Women*, p. 204.
58. *Scientific American*, "Clinical Trials Have Far Too Little Racial and Ethnic Diversity," September 1, 2018. www.scientificamerican.com.
59. Quoted in Criado Perez, *Invisible Women*, pp. 222–23.
60. Quoted in Ja'nel Johnson, "For Many Minorities, Chronic Diseases Go Undiagnosed and Untreated," WFPL, August 31, 2015. https://nextlouisville.wfpl.org.
61. Quoted in Kristen Jordan Shamus, "Family Ravaged by Coronavirus Begged for Tests, Hospital Care but Was Repeatedly Denied," *USA Today*, April 20, 2020. www.usatoday.com.
62. Quoted in Carolyn Y. Johnson, "Racial Inequality Even Affects How Long We Wait for the Doctor," *Washington Post*, October 5, 2015. www.washingtonpost.com.
63. Criado Perez, *Invisible Women*, pp. 234–35.

FOR FURTHER RESEARCH

Books

Nicolas Chevassus-au-Louis, *Fraud in the Lab: The High Stakes of Scientific Research*. Cambridge, MA: Harvard University Press, 2019.

Barbara Diggs, *Racial Bias: Is Change Possible?* San Diego, CA: ReferencePoint, 2023.

Caroline Criado Perez, *Invisible Women*. New York: Abrams, 2019.

Jennifer L. Eberhardt, *Biased*. New York: Penguin, 2019.

Stuart Ritchie, *Science Fictions: How Fraud, Bias, Negligence, and Hype Undermine the Search for Truth*. New York: Metropolitan, 2020.

Angela Saini, *Inferior: How Science Got Women Wrong—and the New Research That's Rewriting the Story*. Boston: Beacon, 2017.

Sara Wachter-Boettcher, *Technically Wrong*. New York: Norton, 2017.

Internet Sources

Julia Angwin et al., "Machine Bias," ProPublica, May 23, 2016. www.propublica.org.

Johana Bhuiyan, "Facial Recognition May Help Find Capitol Rioters—but It Could Harm Many Others, Experts Say," *Los Angeles Times*, February 4, 2021. www.latimes.com.

DeNeen L. Brown, "'You've Got Bad Blood': The Horror of the Tuskegee Syphilis Experiment," *Washington Post*, May 16, 2017. www.washingtonpost.com.

Chris McGreal, "Big Oil and Gas Kept a Dirty Secret for Decades. Now They May Pay the Price," *The Guardian* (Manchester, UK), June 30, 2021. www.theguardian.com.

Nature, "Henrietta Lacks: Science Must Right a Historical Wrong," September 1, 2020. www.nature.com.

Ritu Prasad, "Eight Ways the World Is Not Designed for Women," BBC, June 5, 2019. www.bbc.com.

Jonathan D. Quick and Heidi Larson, "The Vaccine-Autism Myth Started 20 Years Ago. Here's Why It Still Endures Today," *Time*, February 28, 2018. https://time.com.

Scientific American, "Clinical Trials Have Far Too Little Racial and Ethnic Diversity," September 1, 2018. www.scientificamerican.com.

Websites

American Civil Liberties Union (ACLU)
www.aclu.org
The ACLU is dedicated to ensuring individual rights and liberties. The group's activities, many of which occur in courtrooms, focus frequently on racism and sexism. Its website includes position papers and information about racism in artificial intelligence.

American Medical Association (AMA)
www.ama-assn.org
The AMA is the nation's largest organization representing physicians. By typing "medical ethics" into the search bar, visitors will be directed to the organization's code of medical ethics and other articles that discuss this topic.

US Food and Drug Administration (FDA)
fda.gov
The FDA is a federal agency responsible for protecting and promoting public health and safety in connection with drugs, medical devices, food, cosmetics, and more. Its website offers information on new drugs and medical devices and the processes they go through in order to be approved, especially as it pertains to human subjects.

Gendered Innovations
http://genderedinnovations.stanford.edu
Gendered Innovations tracks and works for greater inclusion of women in scientific studies. The website includes discussions and examples of its work and the reason for it.

National Highway Traffic Safety Administration (NHTSA)
nhtsa.gov
The NHTSA is a federal transportation safety agency. Its website focuses on vehicle safety, including information about crash test dummies and how they are used to determine how safe a vehicle is, given certain types of crashes.

INDEX

Note: Boldface page numbers indicate illustrations.

algorithms
  in artificial intelligence, 29, 31
  bias in, 29–31
  medical, 54
  proprietary, 31
American Civil Liberties Union (ACLU), 35
artificial intelligence
  algorithms and data as basis of, 29, 31
  bias in, 29–31
  facial recognition software, 34–37, **36**
  theory behind use of, 28
  voice recognition software, 32–34, **33**
Asian Americans, 36
attention deficit disorder, diagnosis of, 53
autism, 19, 53

"bad blood," 9
Bailey, Harry, 16
biases in science
  algorithms and, 29–31
  basis of product design and, 38
  facial recognition software and, 34–37, **36**
  implicit, 28
  long-held, 4–5
  risk assessments for criminal behavior, 31–32
  voice recognition software and, 32–34, **33**
Black Americans
  Bailey mind control experiments on, 16
  biases and prejudice against, 5
    in emergency rooms, 54–56, **55**
    in risk assessments for criminal behavior and, 31–32
    in school discipline, 28, **30**
  and case of Henrietta Lacks, 11–15, **13**, **14**, 16–17
  facial recognition software and, 35, 36
  medical diagnoses for, 53, 54
    use of algorithms in, 54
  numbers of, in most medical research studies, 52
  trust in science and medicine by, 17
  in Tuskegee study, 8–11, **10**, 12, 16–17

body armor, **42**, 42–43
British Medical Journal, 25

cancer
  diagnosis of, 53, 54
  Henrietta Lacks and, 11–15, **13**, **14**
cardiopulmonary resuscitation (CPR), 46
cars, product design of, **39**, 44–47, **45**
cell lines, 13
cell phones, product design of, 39–40
Centers for Disease Control and Prevention (CDC), 18, 48
Central Intelligence Agency (CIA), 16
chicken pox, 18
clinical trials, 48–52, **50**
Clinton, Bill, 11
Coates, Ta-Nehisi, 18
Commission for Racial Equality (Britain), 30
Commonwealth Fund, 18
communality (Mertonian norm), 24
coronary stents, 50–51
COVID-19, 18–19, 54–55
crash test dummies, **45**, 45–46, 47
Criado Perez, Caroline
  on algorithm-driven products and inequality, 29
  on calorie counts on treadmills, 39
  on crash test dummies, 46
  on drugs that work on women but not on men, 52
  on men as default humans in scientific research, 47
  on need for revolution in medicine, 56
  on women in product design, 39
criminal behavior, 31–32, 34–35
culture and science, 4–5

Darwin, Charles, 5, **6**
data
  artificial intelligence and, 29, 31
  fabrication and manipulation of
    Yoshihiro Sato and, 21
    scientific inquiry and, 22
    Andrew Wakefield and, 21–22
  facial recognition software and, 35
disinterestedness (Mertonian norm), 24
drugs, testing, 48–52, **50**

earth science, 4
emergency rooms, racism in treatment in, 54–56, **55**
ethics
    and behavior of scientific researchers, 22
    code for scientific research, 6–7
    described, 6
    and Mertonian norms, 24
    and testing hypotheses, 27
    See also medical ethics
European Journal of Public Health, 25
Exxon Corporation, 26–27

facial recognition software, 34–37, **36**
Farber, Daniel, 26
fossil fuel industry, 25–27, **26**
Fowler, Gary, 54
fraud
    by Yoshihiro Sato, 21
    tobacco companies' studies and, 24–25
    by Andrew Wakefield, 19–22, **20**

Gambrell, Keith, 55
Gey, George, 12–13
Gey, Margaret, 12–13
Green, Alexander, 56
Greenwald, Anthony, 28

Harvard Business Review, 30–31
heart attacks, diagnosing, 52–53
HeLa cell line, 13–14
Heller, Jean, 11
Hoffert, Martin, 27
Holder, Eric, 31
hospitals, racism in treatment in, 54–56, **55**
Human Rights Watch, 37
hypotheses, 27

immortal cells, 13–14
implicit bias, 28
influenza, 18
informed consent, 9–10, 48
Invisible Women (Criado Perez), 56

James, Reginald, 12
Johns Hopkins Hospital (Baltimore), 12
Journal of Health Psychology, 25

Koch, Christina, 41
Kovacevic, Jelena, 31

Lacks, Henrietta, 11–15, **13**, **14**, 16–17
Lancet, 19, 20, 21, 22
life sciences, 4

Lipe, Rebecca, 42–43

Mark, Saralyn, 41
Massachusetts General Hospital, 56
Mayhew, Deanna, **42**
McClain, Anne, 41
McDonald, Rosealine, **42**
measles, mumps, and rubella (MMR) vaccine, 19–22, **20**
Medical Apartheid (Washington), 16
medical devices, testing, 48–52
medical diagnoses, 52–54
medical ethics
    in clinical trials, 48–52, **50**
    and denial of treatments known to be effective, 10–11, 12
    and financial benefits from research, 15
    and informed consent, 9–10, 48
men
    as default humans in scientific research, 47
    car accident injury and survival rate of, 47
    drug trials and, 52
    facial recognition software and, 36
    medical diagnoses for, 53
    product design of items for civilians and, 38, **39**, 39–40
    salaries of women compared to, 28
    superiority biases for, 5
    in Tuskegee study, 8–11, **10**, 12
    voice recognition software for, 33, 34
Merton, Robert, 24
Mertonian norms, 24
military equipment, product design of, 41–43, **42**
mind control experiments, 16
Moore, Steven, 40
Mounts, Jessica, 43, 44

Najibi, Alex, 36–37
National Aeronautics and Space Administration (NASA), 40, 41
New England Journal of Medicine, 22
Northpointe, 31
Norton (computer security company), 35, 36

organized skepticism (Mertonian norm), 24

Page-Gourley, Rebekah, 33
Perelman School of Medicine (University of Pennsylvania), 46
physical sciences, 4
polio, 14
Pollard, Charles, 9

privacy concerns, facial recognition software and, 35–36
product design
    of automotive vehicles, **39**, 44–47, **45**
    basis of, 38
    of cell phones, 39–40
    of military equipment, 41–43, **42**
    of scientific clothing and protective gear, 43–44
    of space suits, 40–41
    of treadmills, 39
ProPublica (website), 31

racism. *See* Black Americans
Reasoner, Harry, 11
risk assessments, 31–32
Ritchie, Stuart, 24
Rivers, Eunice, 12
Ryan, Laurie M., 49

Salk, Jonas, 13–14, **14**
Sato, Yoshihiro, 21
Science Fictions: How Fraud, Bias, Negligence, and Hype Undermine the Search for Truth (Ritchie), 24
sciences, divisions of, 4
Scientific American, 52
scientific research
    data fabrication and manipulation in, 22
    design of clothing and protective gear for, 43–44
    ethical code for, 6–7, 27
    fossil fuel industry and, 25–27, **26**
    men as default humans in, 47
    tobacco companies and, **23**, 23–25
scientific theory, 4
secondhand smoke, 25
sexism. *See* women
Skloot, Rebecca, 14
smallpox, 18
smoking, **23**, 23–25
space suits, product design of, 40–41
syphilis, Tuskegee study of, 8–11, **10**, 12
systemic racism, 54–56, **55**

Tatman, Rachael, 34
Terry, Luther, **23**
thalidomide, 51
Time magazine, 22
tobacco companies, **23**, 23–25
treadmills, design of, 39
Tsongas, Niki, 42
Tulane University, 16

Tuskegee study, 8–11, **10**, 12, 16–17

universalism (Mertonian norm), 24
University of Pennsylvania, 46
US Food and Drug Administration (FDA), 49, 51
Usher, Thomas, 53–54
US Public Health Service (PHS), Tuskegee study, 8–11, **10**, 12, 16–17

vaccines and vaccination
    controversy surrounding, 18–22, **20**
    effectiveness of, 18, 19
    rate of, in United Kingdom, 22
vehicles, product design of, **39**, 44–47, **45**
voice recognition software, 32–34, **33**
Vonderlehr, Raymond, 10

Wakefield, Andrew, 19–22, **20**
Washington, Harriet A., 16
Weinberg, Steven, 4
White Americans
    facial recognition software and, 35, 36
    inaccurate risk assessments for criminal behavior and, 31–32
    medical diagnoses for, 53, 54
    and treatment in emergency rooms, 54–56, **55**
    use of algorithms in, 54
    school discipline disparities and, 28, **30**
    superiority biases for, 5
Wild, Sarah, 53
Womanikin, 46
women
    algorithms biased against, 30, 31
    car accident injury and survival rate of, 47
    of childbearing age and clinical trials, 51
    drug trials and, 52
    facial recognition software and, 36
    inferiority biases for, 5
    medical diagnoses for, 53
    men as default humans in scientific research and effect on, 47
    numbers of, in most medical research studies, 49, 51
    performance of CPR on, 46
    product design of items for civilians and, 38–40, **39**
    product design of military equipment and, 41–43, **42**
    salaries of men compared to, 28
    voice recognition software for, 33, 34

Zucker, Irving, 51